A Grateful Refugee Kid's Recollections

By
John Larry Baer

(formerly known as)
Hans Lothar Baer or "Frechdachs"

PublishAmerica
Baltimore

© 2010 by John Larry Baer.
All rights reserved. No part of this book may be reproduced, stored in a retrieval system or transmitted in any form or by any means without the prior written permission of the publishers, except by a reviewer who may quote brief passages in a review to be printed in a newspaper, magazine or journal.

First printing

PublishAmerica has allowed this work to remain exactly as the author intended, verbatim, without editorial input.

Hardcover 978-1-4512-9963-2
Softcover 978-1-4489-4099-8
PUBLISHED BY PUBLISHAMERICA, LLLP
www.publishamerica.com
Baltimore

Printed in the United States of America

INTRODUCTION

In one of my childhood playbooks, Max und Moritz, after their latest escapade, there was a note: *"Dieses war der erste Streich, doch der zweite folgt sogleich."* i.e. this was just the first act of mischief-the second follows right away." Happily, my life has been full of little acts of mischief, as you'll see in the some of the pictures from my childhood. I learned early in one of my college philosophy courses that "Happiness is the *Summum Bonum*" and I concluded early on in my career that if work couldn't be fun, it would mess up my life. And my family deserved a happy dad coming home at night.

Recently, a brilliant, creative friend complimented me after reading a draft of these memoires and asked if there was anything I would have done differently in my life or if there was some big thing that I felt I missed doing and would still like to accomplish. My instant answer was "Not a damn thing!"

You see-it's been a very good life, especially when I consider that if it hadn't been for the Nazis, I would never have had the benefit of getting to Amerike as a refugee kid, nor to have benefited from America's largesse in terms of giving me a wonderful (and free) education, a meaningful career, in which I think I did some good, and in my second career, as an international consultant, to roam the world with my good wife and enjoy oh, so many wonderful vistas and meet so many great people.

As you read ahead, you may find some of the technical stuff, from my 30 year career working for the U.S. Army in manufacturing methods and technology, and my 20 years helping companies around the world improve their productivity, too technical, and maybe even tedious. That's OK-just skip what we learned from some of the best companies in the world in terms of producing superior products, and correspondence and lecture notes from my files that I thought were still relevant today, 30, 40 and 50 years later. It may be more interesting just to read about our escapades as we traveled through Europe, Israel and Asia.

So now that I'm in my third career as a full-time volunteer, pushing wheelchairs at the local hospital, teaching new immigrants "English as a 2nd language", and helping mostly immigrant 2nd graders in our local elementary school how to read, and volunteering at our library and creating and maintaining databases for our Washington National Opera of all the singers, conductors, directors, designers, etc, who've been with them since 1958, I'm still having a ball. When it stops being fun-I'll quit. So, I hope that you enjoy reading about our escapades-and go do likewise.

Contents

Part 1
**Recollections of Hans Lothar Baer-
a Grateful Refugee Kid** 11
  Growing Up in Frankfurt-am-Main 11
  A couple of memories from the Elternhaus in Seligenstadt. 13
  Other Travels in and out of Frankfurt-am-Main 16
  Berthold Baer's Story 17
  November 10th 1938-Kristallnacht and 10 weeks
  in Buchenwald 20
  Papi's Homecoming-50 pounds lighter 22
  Leaving Home and Coming to England 23
  My British-English Education at Lancaster House School 24
  "Die Allerschönste Lengwich" 25
  German Bombing Attacks on Britain 26
  Werner Simon's Story 27
  Life in Amerike 30

Part 2
A Great Education-and it was free! 32
  "Alma Mater, Tech, May She Live on High…" 32
  Sturdy Sons of City College… 34
  The Strike at CCNY 35
  Give 3 Cheers for Old Iowa State, A.M.E.S. I O Way… 35
  My religious experience at Iowa State 37

Part 3
WORK 38
   Learning to make Cortisone at Merck's Rahway, NJ plant 38
   1952-1955 Helping to build the 280mm Atomic Shell at Picatinny Arsenal 38
   Meeting my Bride-to-be 40
   The Honeymoon in Canada 41
   Basic Training at Ft Dix, NJ 42
   "Little John" fuze Project Engineer in Uniform at Frankford Arsenal in Philadelphia 1955 43
   Our first home in Levittown, PA 45
   Chief High Explosives and Pyrotechnics Small Arms Ammunition Section 1957 to 1960 47

Part 4
Life in Maryland 48
   The 1960 Move to Harford County, Maryland 48
   The Harford Jewish Center 49
   Life in Harford County, MD 50
   The Limited War Laboratory (LWL) (1960-1962) 52
   The Munitions Branch-Elmer Landis 53
   The LWL Munitions Branch (1963-1969) 54
   Other Munitions Branch Projects 56
   SASA–the Small Arms Systems Agency (1969-1972) ... 57
   The Demise of SASA & Return to the new LWL (1972-1974) 60
   AMSAA & The Human Engineering Lab (HEL) (1974-1976) 61
   At The Army Materiel Command HQ 1975–1983 62
   The "other" Commands—ARMCOM, AVSCOM, ERADCOM, MICOM, TACOM & the Arsenals and MTAG (Manufacturing Technology Advisory Group) 64

Conferences 64
Recycling Toxic Wastes 66
Educating the Government Manager–1978 66
Another Symposium on Government & The Economy–1980 68
Capitol Hill Workshop–April 1981 68
The Brookings Institution (Williamsburg, VA) 71

Part 5
Our Travels from 1969 to 1983 72
1973 the "first" trip back to Germany & the Altstadt 72
The Train Ride to Venice-Mestre 74
1974 Holy Days in Israel 75
1975 My trip to Korea and Seoul food 76
1976 Berlin and a European CALS Conference 77
1979 Visit to Grand Canyon, Oak Creek Canyon &
Tlaquipaqui 79
Oak Creek Canyon 80
1980 Tour d'Europe encore–France, Italia, Schweiz,
de Nederland 81
Seeing a Bit of Östreich, Deutschland and
de Nederland 86
Miami Beach, FL 86
The 1980 & 1981 California MTAG* Conferences 87
The 13th Annual 1981 MTAG Conference 88
Touring Europe Again in 1982–just for fun 89
Summer of '83 in New England 91
Summer Business in Augsburg, Milan, Zürich, Pilatus,
Luzern, and Bern 92
1983 "Retirement/le Retrêt?"–beginning a new career 93

Part 6
International Management & Engineering Consultants (1983-2003) 123
Ending 1983 at the Institute for Defense Analysis 123
Our first non-business vacation 124
A 10-week Dream Assignment 127
Our "findings" in Sweden and Portugal 132
Findings at SAAB-Scania in Sodertalje, Sweden 133
Findings at Sajo Maskin in Varnamo, Sweden 134
Findings at Volvo Car Corp in Goteborg Sweden 135
Lessons from the International Foundry Congress in Lisbon 137
Renault Machines Outils Fonderies du Poitou 138
Findings at Citroen Usines des Ayvelles, Charleville-Mezieres 139
The GIFA* '84 International Foundry Trade Fair in Germany 139
Findings at the Daimler-Benz Leichmetalgiesserei (light-metal foundry) 140
Findings at Bayerische Motor Werke, München 141
Findings at Urdan & Vulcan Steel and Iron Foundries, Natanya & Haifa, Israel 142
Observations in Hong Kong 143
Findings at the China External Trade Development Council, Taipei, Taiwan 144
And then there's Japan! 146
1985-1987 Proposal Preparation Assistance to IMI (Israel Military Industries) 159
Working at IMI with my spouse 160
1986-My "final consulting assignments" for IMI 161
1985 "Study of the Effects of Foreign Dependencyin US DoD Hardware Procurement" 171
Consulting for Dravo Automation Systems 1987 174

Time out for a trip to Greece, Italy and Germany in 1987 175
DGA International & Oto Melara 1988 to 1989 180
Foster Miller, Waltham, MA-1988 181
1989, a busy year in America and France 181
Going to jail for Black Diamond Enterprises in 1990 186
1991–The Soviet Institute of Metrology and Standardization 188
OBSERVATIONS FROM THE SOVIET UNION 194
Swiss Machine Tool Builders Assn., Zurich 196
Explosives Detection Systems study for our airports (1996) 198
Rio Library InfoTech Conference, Rio de Janeiro, Brasil, 13-14 November 1997 199
The Titan IV "Should Cost" Study at Martin Marietta 200
Titan IV Should Cost—Quality Assurance Recommendation 1 201
Charge to Ballistic Missile Defense Program Managers (Banquet) 11/3/81 202

Part 7
The CALS Program (Computer Aided Acquisition LogisticsSupport around the Globe) 204
First CALS DoD/Industry/NIST Conference April 1989, Gaithersburg, MD 204
CALS Expo '92, San Diego Convention Center, December 1992 205
CALS Europe Berlin, September 1993 206
Korean Management Advisory Council Conference 29 April 1996 207
CALS Pacific '96–Seoul, Korea, 3-6 September 1996 208

12th International Logistics Conference, Athens/Glyfada, Greece, 27-29 September 1996 208
Global Electronic Commerce & Money Management, Athens, Je '98 209
CALS Expo '96, Long Beach Convention Center, Long Beach, CA October '96 212
Mediterranean CALS '97, Istanbul, Turkey 213
CALS Europe '97 Frankfurt-am-Main, Germany 1-2 Oktober 1997 217
CALS & Electronic Commerce 217
21st Century Commerce & CALS EXPO USA 1997 217
GIC France Guidelines for a new European CALS/ ELECTRONIC COMMERCE 218
IV Simposium Industrial "Ingeniera de Vanguardia" 219
Congreso Sostenibles en la Industria de Alimentos, San Jose, Costa Rica, Julio 1999 Keynote 232
1er Congreso Nacional de Ciencia y Tecnología de Alimentos, San Jose, Costa Rica, 19 julio de 1999 233
Helping Siemens transition into Y2K–July 1999 239

Part 8
Conclusions and "Lessons Learned" 241

Part 1
Recollections of Hans Lothar Baer-a Grateful Refugee Kid

Growing Up in Frankfurt-am-Main

Wie fröhlich bin ich aufgewacht
Wie hab ich geschlafen so sanft die Nacht
Hab Dank Du Vater im Himmel mein
Das Du hast wollen bei mir sein.
Behüte uns auch diesen Tag
Das uns kein Leid geschehen mag. *

That was my morning prayer as a child, and something that has stuck with me over the years as a rather satisfying way of starting the day. Strange how 70 years or so later, things like this stick in your mind.

Since those early days in Nazi Germany, and our blessed escape to these glorious United States, I've also learned, mostly from my wife, that "Life is a Crap Shoot"–but we can change the odds, often doing "what's right" and "doing well" while "Doing Good!" As my late mother used to tell me: *"Wie man in den Wald ruft, so schallt es heraus."* (as you call into the woods, so it will echo out.)

It was actually many, many years after the fact, after we had happily escaped Germany in 1939, that I realized how difficult life had become for the Jews in Frankfurt-am-Main. My parents had had a comfortable life, an ample apartment, or so it seemed to my young eyes, on the 2nd

floor of Obernweg 50, where they had moved from Melemstrasse after I was born.

I still remember that, as you entered the spacious hall with its wicker chairs, there was the *Herren Zimmer* (living room) on the left, connected on one side via pocket doors to a large drawing room (later to be occupied by my Onkel Werner,) and on the other side to the large dining room, with its large *Vitrine* and a Viktrola in the corner (where my Omi, Clara Simon, would sleep in days to come.)

Turning right again, another pocket door led to the master bedroom, with its small balcony overlooking the backyard and the large private Rothschild residence (more about that later). The adjoining bath connected to a smaller guest room, later to serve as the "home" of Herr Doktor and Frau Fürst, with its window looking into the "Winter Garten"–a glass enclosed verandah. Across the hall was the smaller maid's room.

* *How happily I awoke; how softly did I sleep this night. My thanks, father in heaven, that you have stood by my side. Protect us also this day that no harm may befall us.*

If you turned right as you entered the front hall, there was a small bathroom and the kitchen with its pantry. And, oh yes, there was also a balcony in front of the *Herren Zimmer*, overlooking Obernweg and later Nazi Swastika flags flying next door. (see photo of me and cousins Peter & Mucki Euphrat in 1936 & 1938 on page 96.)

My father was the manager of Heidingsfelder & Cie, a small brokerage house near the *Hauptwache*–an easy 8 to 10 block walk from our home, down the *Eschenheimer Landstrasse*, past *Eschenheimer Tor*, or by *Trambahn* or streetcar.

On Simchas Torah we would walk to the Unterlindau Sinagoge or the one on Freiherr von Stein Strasse, stopping along the way to pick up shiny chestnuts, newly fallen from trees lining the streets, which I would stuff into my bulging pockets. On holidays we would take a train to Seligenstadt, the village where my father had been born and raised, not far from the Hessian town of Hanau. The *Elternhaus at Steinheimerstrasse* 10, still stands on the corner (1938 family photo) just a block from the *Steinheimer Tor* that once marked one of the entrances into town through the encircling *Stadt Mauer* (City wall.)

Or we would take the tram to Oberursel or *Bad Homburg vor der Höh*, the resort town where my mother was born and her parents owned a small houseware shop, to visit my Onkel Rudolf Haman and Tante Alma. He had been a violinist in the *Kur Orchester*–a gentle man, who was later to die an untimely and unnecessary death at the hands of a drunken GI, who shot him in the stomach.

Onkel Rudolf had married my Omi's sister and lived in a comfortable little cottage on Hasensprung in the middle of a garden, replete with flowers, a Gazebo, a plum and a peach tree, and an apricot tree from which I used to pluck ripe fruit, which I learned many years later, actually belonged to the neighbor.

At other times my parents took me skiing in the nearby Taunus mountains, as soon as I could walk and stand on *"zwoa Bretterl in geführiger Schnee."* Or we would take the train to Friedberg to visit my Tante Else and her large husband, Emil, who ran a butcher shop, or to Eltville on the Rhein to see my Tante Lenchen, her daughter Ella and her husband, Julius, who ran a shoe store together there. I remember well the wonderful smell of the shoe leather, a smell I still enjoy to this day. Funny how we remember people and places by their scent.

A couple of memories from the *Elternhaus* in Seligenstadt.

I remember as a very young and very inquisitive boy, released from the white gloves and neat suit my mother had obliged me to wear, running around on the stone floor from the kitchen with its usually steamy, tasty smells, back along the hall to investigate the pump under the back sink where you came in from the farm yard, and how I scared the hell out of everybody by nearly falling down the well. Then there was that wonderful backyard, which you had to navigate to get to the outhouse (except at night when chamber pots were kept under the bed), past the *"Mist Haufen"* (dung heap)–what we would call a compost heap today, composed of hay from the barn mixed with animal droppings.

In the back on the left was Lotte's stall (the horse used to pull the plow or wagon) and on the right the cow barn. The chickens roamed the yard, except when Hexchen, the family Dachshund chased them. And sometimes we would go out the huge back doors and walk down to the Main, the river which we still visit from time to time when we go back to see my cousin Hans and his wife Marianne who live a five minute walk from the (since-sold and carefully restored) *Elternhaus*.

(A note here, that town rules require that any of the old Fachwerkhäuser, with their exposed wooden beams, be restored approximately to their original design, before the new owner is allowed to renovate the inside.) In this case, the stone stairs just inside the hall on the left as you came in from the front door, which led down to the cold cellar, repository for winter beets, apples and other goodies, were kept intact, but with a modern railing leading down to a neat apartment.

To the left was the family room, where we used to gather and where, one day in my haste to get to the window, I tripped over my grandfather, Julius' leg, (on the left) cracked my head on the window sill and, to my mother's screaming horror, bled like a pig. I recovered, but still carry the scar on my forehead today. To the right was the *"Gute Stub"* and *Speisezimmer* reserved for special meals, and in the *Me & my Opa Julius 1929* back the master bedroom, taken over by Tante Irene, the youngest daughter, who had married Adolf Thoma, a good Catholic farmer, and bore him two sons, Hans and Erich, my younger cousins.

Ernst was the youngest of 14 children, starting with Isidor, who was born in 1885 when my grandmother, Esther (nee Emma Bender) was 25 and my Opa, Julius Baer was 27. He died in 1913. Then there was a baby girl who died in infancy, then Lenchen, born in 1887, who died in 1943 in Theresienstadt; Tanta Lenchen married Julius Simons, and had a daughter, Ella in 1911. They owned a leather store in Eltville on the Rhein–a place I loved to visit, because of the wonderful smell– which I still love to this day. Else, born in 1888, married Emil Adler, and they had a butcher shop in Friedberg-they died in 1943 in Auschwitz.

Then there was Ludwig, born and died in 1890, Benno, born in 1893, who died in the war in 1917; and then my Dad, Berthold, my hero, born in 1894, who died in Levittown, PA just short of his 70th birthday, from emphysema and lung cancer attributed to his many years of smoking and exposure to fumes from the smelters he worked on in later life. One of the many things I admired about my father, beside his normally calm demeanor (in contrast to my mother's frivolous nature,) was that here was a man who had a distinguished desk job as B*ank Prokurist*, dealing on the then, most important Frankfurt stock exchange, who put it all behind him and worked as a scrap metal sorter here in America to support his family.

Betty and Arthur, born in 1895 and 1896, were two more infant deaths. Onkel Robert was born in 1898–the serious uncle, who married Tante Minna Fromm, a devoted, but somewhat, shall we say unusual person. They opted to go East in 1937 via Russia and China to get to San Francisco and wind up with the rest of the family in New Jersey, where she had a lobotomy.

Tante Irene was next, born in 1899, and torn untimely from her family by the Gestapo in 1940, to be transported via Offenbach (from where she sent home a postcard, the first letters of each line spelling out HUNGER) to Auschwitz, where she perished in 1943. She had married Adolf Thoma, a good Catholic, which saved their two boys, Hans and Erich. Tante Setta was born in 1901, married Ludt Baer late in life, and died walking on the street in Queens, NY at the ripe old age of 75. Onkel Willy, the stubborn one, to whom was attributed a slogan *"Ich habe immer Recht,"* (I'm always right) was born in 1903. He too emigrated to the US with his first wife, Hilde Goldschmidt from *Konstanz*, who died from cancer in 1941. He then married Doris Aufseeser, a widow, with whom he had a daughter, Evelyn; they resided in North Bergen until his death in 1968. At her death in 2006, Tante Doris was still a vital 94-year old.

The youngest, Onkel Ernst, born in 1904, married Theresia Adler from *Konstanz am Bodensee*. He too came to New Jersey in 1938, worked at his profession as a baker in West New York, NJ, and raised two daughters, Inge & Irene–the successful cousins whom we still see frequently today. He died in 1981.

Im Seligenstädter Familienhaus

What I remember most vividly was that when we stayed over on a cold winter night, Tante Irene took a hot brick from the fireplace, carefully wrapped in a cloth, and placed it at the foot of my very cold bed upstairs, before I went to sleep.

What I never experienced there, but heard about many times, was that when there was a thunderstorm, my grandmother, Esther (called Emma), would gather the family in the family room downstairs and read from the bible, till the storm passed. Religious? Not really. The custom grew from the need to be prepared in case a lightning strike ignited the hay in the barnyard or struck the house, so folks could put out the fire quickly before the damage got too bad.

Other Travels in and out of Frankfurt-am-Main

I remember going to the Bäckerei on Öderweg with my father on Saturday mornings to buy Brot Aufschnitt of assorted breads and then to the Metzgerei for Aufschnitt of assorted cold cuts. I remember particularly on a Mother's Day, when I was 5 or 6, my father taking me on the child seat on his bike, and going into the woods near Adickesalee to pick wild flowers. On the way home, he composed and taught me to recite to my mother, the following poem:

"Blumen sagt man bringen Glück
drum kehr ich aus dem Wald zurück
mit den schönsten Blümchen die ich fand
und Dir in ein kleines Strässlein band.
Der Muttertag ist Anlass mir
Zu schenken diese Blumen Dir,
Mit dem Versprechen mein liebes Mütterlein,
Ich werd Dir ewig dankbar sein." *

(Flowers are said to bring luck; hence I've returned from the woods

with the prettiest flowers I could find, and bound into a little bouquet for you.
Mother's Day is my reason to give you these flowers,
with the promise, my little mother, I will always be grateful to you.)

As I approached school age I was outfitted in my best shorts and shirt and presented with a *Schul Tüte*–a large conical bag filled with goodies, sweets and my first school supplies *(see photo)* as I was enrolled in the *Holzhausen Schule*. Later, after graduation from this elementary school, or *Vorschule*, like most Jewish children in the area I attended the Philantropin for about a year, before the Nazis unceremoniously closed it in 1938.

One of the "poems' my mother taught me, was:
Vom Vater hab ich die Statur, des Leben's ernstes Führen.
Von Mütterlein, die Frohnatur, die Lust zum fabulieren.
(from father I have stature, life's important tasks;
from little mother, happy nature, the desire to have fun.)

Funny how naïve children can be–one day I went for a walk with my Omi, and when we got to the Eschersheimerlandstrasse, there was our local cop, whom we all knew, and, having heard everyone else do it, I said a friendly "Heil Hitler."

My grandmother nearly dropped in her tracks and when she got me home, explained to me in no uncertain terms that this was an absolute No-No. You learn. Life was relatively peaceful (as far as I could observe, naively), but things were about to change drastically on that momentous day–November 10th, 1938!

(For a better perspective, you may want to take a look at the abbreviated Timeline of Hitler's rise to power in 1933 to his demise in 1945 in the Appendix, and the parallel ascent of the Nazional Sozialistische Arbeiter Partei, the National Socialist Worker's *Party*

Berthold Baer's Story

It was midday, or close to it, on a chilly November 10th in 1938. Your Opa, my Papi (as I called him,) was due home for lunch from his job as a stock broker at Heidingsfelder & Cie, on Steinweg in Frankfurt am Main. Two men had come to our 2nd floor apartment at Oberweg 50 just a little while before. One was Herr Muller, our genial policeman from the neighborhood police station, whom we knew well and even liked. (We called him a *Schupo*, short for *Schutz-polizei.*) The other was obviously a Gestapo man-albeit in mufti. Very polite and considerate under the circumstances. They waited for Papi in the Herren Zimmer-our little living room.

As I said before, we had a rather spacious flat-a large foyer, replete with wicker furniture where you could rest or play on a rainy day. In front on the left was the room occupied by my Onkel Werner. He was a lawyer and full of fun.

He always sent me funny letters with envelopes on which he drew pictures of the trail the letter had to take to get to me and balloons around the stamps. (*if you're interested, send me an e-mail and I can show you what they looked like.)*

Even when he got caught in a blizzard skiing on the Matterhorn and lost his thumb to frostbite, he was ever the happy fella. Mutti loved him very much-he was her only brother and when he died at Auschwitz she was heartbroken. Never wanted to go back to Germany after that. But I digress.

Onkel Werner's room, the Herrenzimmer and the adjacent dining room-all connected by huge (to me) pocket doors, all faced the street, each room with its own little balcony. At some point, since my Omi (Mutti's mother) had moved in with us and slept on a couch in the dining room, they had put a couch in front of the connecting door. What a challenge this was for me, a 10-year-old rascal, too often constrained in white gloves and a proper suit with gray flannel shorts. I liked to slide open the pocket doors and jump over the couch-except that one day I tripped and fell flat on my stomach. Boy, were those ladies hysterical-but it just knocked the wind out of me and I was fine in a few minutes.

As to the balcony, being good German Jews, with the accent on the German, we flew the flag on appropriate holidays. Just one small problem there-a lot of the neighbors were flying the Swastika emblem and that bothered Onkel Werner and my Papi a lot. What did I know of all this?

A GRATEFUL REFUGEE KID'S RECOLLECTIONS

Oh yes, I still have a photograph of me with my favorite cousins, Peter and Mucki Euphrat (his real name was David, but only his wife, Miriam, called him that and, of course it appeared on his tombstone.) The photo, in which we all have mustaches drawn on our faces with a burnt cork (as we used to do things just for fun in those days) clearly shows a Nazi flags jutting from adjacent houses.

From the dining room, with its large black ebony table with beautifully carved heavy legs and its two sideboards or Vitrines, one door led out into the foyer and another set of pocket doors to the adjacent master bedroom occupied by my parents. European bedrooms weren't fitted with walk-in closets the way we have them. Instead one had to buy a *Schrank*, a huge wardrobe to hang suits and dresses, with drawers for shirts and socks and underthings. There was a small balcony there too which looked out over the tree studded back yard and one of the smaller Rothschild mansions (more of that later.)

(By the way, although Oberweg 50 had been destroyed by bombs during the war, the backyard looked exactly in 2006, as I remembered it from 1938.)

That bedroom connected to a bathroom, whose other door connected to a small room that used to be mine-till my folks rented it to a Jewish couple, Dr. of Jurisprudence and Mrs. Fürst. I was too young to be told why or how they lost their own residence. Suffice it to say that my folks took them in.

Their room had just one window, which faced into the rather large glass enclosed "Winter Garten." This was another neat place to play when the weather was inclement since my meticulous mother didn't like me to play in the mud, as I would have preferred. A door connected the Winter Garten to a room I shared with our *Dienstmädchen* or housemaid. The only other room, which abutted this one and had a door to the lobby, was the comprehensive kitchen, with its gas stove, *Eis Schrank* (an ice box) and a real pantry facing an outside wall, to store things.

(We also had Sinsel, a lady who came in to do sewing and one to iron.) Oh yes, there was a small Toilette between the kitchen and the front door of the flat.

So there you have the setting on this momentous day, which was to end in what was later to be known as *Kristallnacht*, the night of smashed crystal ware, so common in "comfortable" Jewish homes.

November 10th 1938-Kristallnacht and 10 weeks in Buchenwald

At the usual appointed time my Papi came home for lunch, to be told in hushed tones about the "visitors" in the front room. He greeted them and they politely told him to go ahead and have his lunch and then they would talk. At this point no one except me had any appetite-typical of a 10-year old boy.

When my dad met with the gentlemen, they politely suggested that he pack a small suitcase with some warm clothes and to be sure to take his citation for having been a World War I "front soldier" and his iron cross-a touch that was to save his life. Anyway, he did what they suggested and left with them for a "short trip", leaving the rest of the family in tears and very frightened.

That night I remember my Mutti let me sleep in her bed and I also remember that the wind howled that night as I had never heard it before. Around midnight we heard windows being smashed in the Rothschild house across the way in back, screams, more systematic smashing and then a deadly stillness. And thus I finally fell into a deep exhausted sleep.

The next weeks were a period of anguish for my mom and, to a lesser extent, for me, because I really didn't know what was going on. Dad called home once to say he was all right. The next call came six long weeks later from a place called Buchenwald! (about 4 hours to the North by train.)

Much later we were to learn that my dad had spent the first night of his civil arrest with several thousand other adult Jewish males, scrubbing the marble floor of the conference center with their toothbrushes. The next day they were taken by train to Buchenwald, where they were herded into unheated barracks with stacked bunks.

Periodically they were chased out into the cold by the Kapos, the political prisoners, who were their designated overlords. There they were made to stand at attention while guards with snarling German Shepherd watchdogs, were supervised by the equally menacing and infamous Ilse Koch.

It was only many years later that dad told me the tales of how "the bitch of Buchenwald" searched for men with interesting tattoos which she had peeled off to make into lamp shades. And how when some of the weaker men dropped in their tracks, the guards sicked the dogs on them. Not a pretty sight. And periodically someone who fought back was shot in cold blood, and occasionally they would hang an especially obnoxious prisoner, whether civil (the Jews who had been rounded up) or a homosexual prisoner, tagged with a pink armband, or even a communist, arrested as political prisoners. These later also served as prison doctors, if they were so licensed in the outside world.

What with the bitter cold and lack of decent food-how much nourishment can you get from a thin gruel?-Dad got weaker and came down with pneumonia. In fact, we got word from a friend who had been released early in December, that he had seen dad on a stretcher carried over to the infirmary just before he left. Dad was also to tell us later that when the camp doctor used a tongue depressor on his ultra-sensitive palate, dad threw up his gruel over the doctor. That "good man" pulled his pistol-and then thought better of it and let my dad live.

Meanwhile, he sent us a postcard with the following note:

I have, until further notice, mail restriction, can, therefore neither send nor receive letters, cards or packages. Inquiries to the commander of the camp are forbidden and will extend the mail restriction.

Six weeks after entering Buchenwald, Dad was once more being shlepped to the infirmary when he heard the announcement over the camp's loudspeaker that "Alle Schwein Juden" (all the Jew pigs) who had been front soldiers in the war were to come to the front office WITH their papers and their belongings.

The rest is history. The Kapo clerk looked at the documentation and, satisfied with what he saw, said the magic words "Raus!" or "get out."

Dad could hardly believe his good luck, but asked no questions. He hobbled as best as he could to the train station and called us to say that he would be home as soon he could get a train. .

Papi's Homecoming-50 pounds lighter

And thus it was that he blessedly returned home to a tearful reunion with us, and my parents made plans to leave Germany as quickly as possible. Dad had been persuaded only early in 1938 to apply for a visa to the United States. Other friends from his card group (they played *Skat* once a week on Saturday night) had already left for America and had urged him to at least request a Visa. Dad, the prototypical, "honorable German Jew" always felt that since we were Germans (albeit of Jewish conviction) we were fairly secure and all the Nazi nonsense would soon be contained.

It was only when the Nazi goons, the SA brown shirts and the *Schutzstaffel* (SS) black shirts started to smash people's heads along with their store fronts and homes, that he began to feel vulnerable and that maybe it was time to get out. Smart choice, albeit a bit late.

The Nazis had already "bought" our delicate, hammered silver baskets and decorations for 1 Mark a pound! So my folks packed whatever we could send to America in the way of furniture and household goods. Some of these paintings, bric-a-brac, utensils and even kitchenware are heirlooms today, conserved in our home with pride.

(A letter to the SS, like the postcard & pictures of my dad before and after Buchenwald, that enclosed certification from the British consul to permit immediate emigration from Germany to England, are in the middle of this book.) But here, is the translation:

To the Commandant of Concentration Camp Weimar-Buchenwald, 28 November 1938

Regarding: Imprisonment of Berthold Baer

In the enclosure I am sending a copy of my letter to the Secret State Police Frankfurt, requesting the release of my husband, who has been imprisoned at the concentration camp Weimar-Buchenwald. I ask that this request will be honored as soon as possible.

Leaving Home and Coming to England

Over the next few weeks we packed clothes and whatever we thought we could carry with us, including Dad's stamp collection, and boarded a train for Holland. One of the things that helped to speed us out of Germany were posters like the warning to "good" Germans not to shop at Jewish stores (the few that hadn't had their windows smashed and contents looted.)

At the border check point, the Nazi customs official carefully thumbed through said collection and brazenly "helped himself" to a few dozen of the finest, most valuable stamps. Under the circumstances we were just glad when, after very careful and probing scrutiny, he closed the suitcases and stamped our passports, permitting us to re-board the train to continue our journey.

It was only after we had crossed the border into "freedom" that my parents heaved a joyful sigh of relief. We were headed for Vlissingen to catch the ferry that crossed the English Channel to Harwich in England and a temporary safe haven, providentially and graciously provided by one of my Dad's cousins in London's Golder's Green section. Herman Baer and his wife Trude and their son Harold (a reclusive Veterinarian in Toronto, since deceased) welcomed us and we soon grew accustomed to life among English Jews who tend to flock to this part of town. I always remember arriving in the British train station and hearing people "laugh in a strange language." I knew no English, but like many a refuge kid welcomed into the UK, I soon got the hang of this new and strange mother tongue spoken around the world.

We learned to greet the milkman who delivered milk products and fresh milk and cream daily at our kitchen door, trudging up the outside iron staircases with his load of bottle filled with luxurious nectar and taking away the empties. We took walks to Chomley Gardens and took the bus to Cricklewood, and, before long met HMBEJP Francisca Gluckstein. HMBEJP stands for Her Majesty's British English Justice of the Peace. Her husband was a barrister in the House of Commons, and she felt constrained to try to help some of these poor refugees.

She took us under wing, and besides helping my parents with those routine things required when residing in a new country, decided that my name "Hans Lothar" just wasn't going to be proper in this new land. She translated Hans to John and decided the closest thing to Lothar was Larry. And so I got my new name–John Larry Baer.

My British-English Education at Lancaster House School

Meantime, Mr. Areen Paul Grundy, the headmaster of Lancaster House School in Action SW3, and his charming sister, Ian, decided to take in two refugee boys–one, Joe Taglicht, separated from his parents in Lodz, Poland, and me. Thus began my education as a proper English student. No, I was not illiterate. I had done rather well in the German educational system, first at the Holzhausen Schule, a public school a short walk from our flat, and then, as the Nazis restricted our attendance as Jews in Aryan schools, the very prestigious all-Jewish Philantropin. I was rather good in math, but not always so good in deportment. It hasn't changed.

So while my parents moved from the Baer's Golders Green flat to serve as caretakers of the home evacuated by some British Jews who opted for the safety of America's shores, I attended class briefly at a London County Council School, before being welcomed into this proper British boarding school-complete with the unwanted homosexual approaches by some of the older students in the dorms.

Along with Joe Taglicht, now a Professor at Tel Aviv University, I learned my English, math, French, Latin and later Greek. Then, as the war with Germany went into high gear and the bombs began to rain on London, they wisely moved the school to a country estate in Shiplake, near Henley on Thames. In the summer, before classes began each morning we would run through the meadow down to the river with Grunter (our nickname for the headmaster) for a chill swim in the Thames.

During the summer I also learned how to pole a punt along the river and to pump the organ in the nearby church. Joe and I would take turns pumping the organ since neither of us was supposed to participate in a Christian service. The fact that we occasionally over-pumped and made the organ squeak, earned us a sharp reprimand from the organist. Still, it was always a nice walk through the woods to this prototypical little country church-especially in the winter snows.

In September of 1940 I was finally recalled to London to join my parents. Communication was a problem since they still conversed mainly in German and broken English, and I was pretty deep into English. But then what children and parents don't have a communication problem.

"Die Allerschönste Lengwich"

Here, I must make a quick aside to tell you about the language used by the German refugees. It was so prevalent that indeed a book was published in England, entitled "Die *Allerschönste Lengwich*" replete with stories such as the following.

"A refugee woman gets on one of London's ubiquitous double-decker buses and gets into an argument with her *hoosband*, who wants to ride upstairs, while she is afraid of climbing the narrow steps as the bus bounces along. So she opts to sit downstairs, while the husband goes up to enjoy the view from the top deck.

Soon the ticket taker comes along, calling out "tickets, tickets, please." When he gets to the lady, she very properly informs him: "*the lord is above.*" To which he replies: "that I know, laidy, but 'e won't pay your fare for you."

Then there was the man renting a room, who was chilled by the frugal Brit's lack of heat, who wrote to his landlady, carefully translating German idiom into what he believed to be proper English:

"My most expensive lady *(meine teuerste Dame)*. There is a train *(Zug, meaning draft)* in my room, and if I don't get another ceiling *(Decke or blanket)*, I undress *(ziehe aus, not ausziehen move out)* tomorrow."

And finally the two refugees greeting each other upon meeting, one asking the other: "Ken you Greengroft Gardens?" to which the other replies: "And if I can." *(und ob ich kann.)* As I said, there's a whole book of such broken English.

Or one refugee asking another: *How much vatch?"* and the other, pointing to his watch, replying: *"such much."*

German Bombing Attacks on Britain

Incendiary bombs falling throughout town, even in our backyard, caused most people to spend their nights in subway stations or, as in our case, huddled into a large, centrally located room on the first floor of the home my folks were protecting-now filled with about a dozen stray folks taken in for shelter.

Then came the blessed day our Visa number for entry to America had come up! We packed our belongings and took a train for Liverpool, to embark on the HMS Scythia, a small steamer, which took the last load of children to safety.

It wasn't easy. On the night before our scheduled departure, escorted by British destroyers for protection against marauding German submarines, our Cunard Star liner took a glancing blow from a German bomb that knocked a plate out of our ship's hull. It was quickly patched up the next day, and the convoy sailed for America.

The trip, for a 12-year-old boy, who didn't know enough to be scared, was a bit of a lark. It culminated in one of THE highlights of my long, blessed and wonderful life when on October 3rd, 1940, the first day of Rosh Hashanah, our New Year, we saw the lights of Long Island and sailed past the Statue of Liberty.

Those lights, after London's depressing blackout of the previous many months, were one of the most welcome sights for this young refugee kid. A sight always to be remembered! From this point in time life for the Baer's had nowhere to go but up. As we became acclimated, we were to learn, first hand, that America is truly the land of opportunity.

Like most immigrants to America my folks did their best to learn English, even though, like many refugees, be they Italian, Greek, Russian or German, they fell back into their native language at home, and often in public. Regrettably, in recent times, while virtually all Asian immigrants, like the Europeans before them, have strived mightily to learn English, far too many of the *Hispanics* in the current influx, have made no effort to learn English, and expect the rest of Americans to "push 1 for English, *por espanol indice dos.*" How sad for America! Of course, Those Spanish speaking immigrants who have opted to ignore this basic wisdom, are doomed to remain 2nd class citizens and become a burden on the state.

Werner Simon's Story

Before we make the transition to our new life in America, let me interject a few lines about my Onkel Werner, who sent me hand-decorated envelopes from Berlin, where he was the Director of ORT, Organization Rehabilitation et Travail (or Training.)

ORT's purpose was to train Jewish merchants, salesmen and bankers in useful handicrafts to include farming and how to operate machine tools.

I don't know too much about the operations of ORT in Berlin, although some of his students and teachers have survived and still live in England. What I did find out was that my uncle had to deal with Eichmann–head of the Gestapo's "Office for Jewish Emigration". Ah yes, when in hell, the devil is the highest society, and you have to deal with the devil to get anything done. In this case, that meant negotiating with Adolf Eichmann to obtain exit visas for ORT's students and teachers to be transferred to an ORT training facility in England (which did not, in fact, exist.) It worked, and most of the staff were able to emigrate.

Until shortly before the Second World War, the students and teachers attended the ORT Technical Engineering School in Berlin. Fearful of what lay ahead, leaders of ORT managed to convince the Nazi authorities that 102 students aged 16 to 17 together with six teachers and their wives should be sent to England to continue their studies at a similar ORT school in Leeds.

In fact, at that time no such school in Leeds existed. Taken in by this deception, the Nazi authorities agreed and Adolf Eichman, eventually responsible for transporting millions of Jews to the concentration camps, signed a letter permitting them to leave Germany (the original letter remains today in the archive at World ORT Union's offices in London).

On 27 August 1939, just one week before England declared war, the students, their teachers and the teacher's wives, boarded a train in Charlottenburg, Germany, eventually reaching England via Holland.

Once in England, British ORT, under the leadership of Lt. Col. J.H. Levey, made arrangements for a school to be prepared. They were finally able to continue their studies at the ORT Technical Engineering School in Rosville Avenue, Leeds. ORT Berlin was closed by the German authorities on 21. März 1940, and a year later Adolf Eichman absorbed ORT schools into the "Jewish Department" of the German Government.

Werner Simon's Obituary

WERNER SIMON the young Frankfurt lawyer, who never lost the accent of his homeland dialect, was born in Bad Homburg on 10 July 1903. His experience in the Jewish youth movement as a member of the "comrades"–crucially shaped his nature and life's course. As a result his life was devoted to beautiful things in art, music and imagery. He was equally devoted to working with youth groups and sports. He was a passionate skier and sailor. After he lost his job due to National Socialist legislation, he was appointed in 1934 to Berlin to the management of the Federation of Jewish front soldiers.

In the later 30's he transferred first the management of the Jewish liberal newspaper and, after their deactivation, the administration of the local ORT school to Berlin, which was closed by the Gestapo in 1941. Due entirely to his initiative and energy he made it possible for most of the pupils and teachers of this school to emigrate to England before the outbreak of war and thereby saved their life.

He himself remained behind, because he believed that he was responsible for the continuation of the ORT school. This devotion cost him and his family their life. He came to Theresienstadt on 16 July 1943 with his young wife, Eva nee Blumenthal and his son Dan, who was born in Berlin in 1940. It was laconically reported in an official report from 1953 that "His further whereabouts could not be determined."

(Werner Simon, his wife and son, were in fact deported to Auschwitz, where they met their death, according to my grandmother, Clara Simon, who was in Theresienstadt with them, and refused the offer to "go East for a better life." She was rescued by American troops several weeks later, and after recuperating in Switzerland, came to the United States, to live with us for a few months, before she succumbed.)

Werner Simon, director of the ORT school in Berlin who negotiated with the German authorities, stayed behind. Sadly, he and his family perished at the hands of the Nazis. Why my uncle, his wife, Evi and their little boy, Dani, did not go along I was never able to find out. Suffice it to say that in 1942 they, along with my grandmother, Clara Simon, were shipped to Theresienstadt (*Terezin*) in Czechoslovakia. Terezin was a show place among camps, often visited and photographed by the Red Cross, showing Jewish internees playing in an orchestra, doing stage productions and seen strolling the streets of this Potemkin Village of handsome facades, hiding the hovels behind them. (the last picture I have of him shows Werner in happier times in Berlin in 1941, with his wife Evi and their boy, Dani, who perished at age 4 in Auschwitz.)

As a "senior person of importance" Werner Simon was put in charge of bread rationing, but was soon convinced that the situation and food supply were better "in the East." Accordingly, he and his family got on the train, for what was to be his final trip to the chimneys of Auschwitz. My grandmother refused to go along and a few weeks later was liberated by American troops, and after a period of rehabilitation and recovery in Switzerland, was able to come to America, where after just a few months, she passed away.

Life in *Amerike*

So here was my father, educated, with a prominent position in Frankfurt, reduced, like so many refugees, to trying to sell vacuum cleaners, while my mother kept house. We were living in a German-Jewish enclave in Washington Heights, at 177th Street and Broadway in Upper Manhattan, unlike the non-Jewish Germans gathered in shops and apartments in Yorkville in mid-town Manhattan.

One day, my mother went to a butcher shop on Broadway and made a modest purchase, but, in trying to make the best buy with still broken English, she left her purse on the counter with about $100–our entire capital–in it.

As soon as she got home, she realized to her horror, what she had done and ran back to the store. Unhappily, someone who must have needed the money as well, had picked up the purse and there went our total reserve. Fortunately, my father's friend, Karl Stoll, his *Skat* partner from Germany, who had been our generous Visa guarantor, came through again and hired my dad, once a stock broker, to work in his factory, Stoll Metal Corporation, as a scrap metal sorter!

What a comedown–and yet, he took it in his stride, coming home to our apartment late each evening, dead tired and dirty. By then we had moved from 171st Street in Washington Heights, to a brownstone on Willow Street, near the well-known Penny Bridge in Brooklyn, where my mother, the "lady" who had gone to "finishing school"–held the job of building super.

She too, remarkably, took this comedown in stride. [Their attitude has made a lifelong impression on me, and prompted me to take life's petty upsets in stride, and, hopefully, to be a better person, giving a helping hand to those in need.]

I walked each day to PS 29, the nearest Public School, attended largely by Italian kids from Red Hook, from whom I quickly learned all the appropriate Italian cuss words and inappropriate hand and arm signals. It was not much of an education, quite a comedown for me, as well, from PS 177 in Manhattan, and certainly from the private school I attended in Acton–more in line with the London County Council School I had attended briefly, before Arreen Paul Grundy and his sister, Ian, took me in to the school they ran.

Part 2

A Great Education-and it was free!

"Alma Mater, Tech, May She Live on High…"

Still, my education was adequate, so that when I took the entrance exam for Brooklyn Tech, the only High School that I know of, that required an entrance exam, which even the elite Stuyvesant HS and Bronx HS of Science, did not demand, I found myself as one of the lucky 1200 students admitted, of which only some 400 were expected to graduate. The rest unceremoniously dropped out and transferred to less demanding schools.

And so I spent four rather happy years at Brooklyn Tech, taking the chemistry course, working at the radio station, WNYE from a small studio on the roof, next to the Hydroponics Club hot-house, learning about Industrial Processes from Mr. Clarence L.E.G. Sjogren (funny how names stick with you over the years,) and Social Studies from Mr. James Harris, one of the few black teachers.

My worst encounter was with Herr Mueser, the head of modern languages, who was upset with me when I got only 96 on the German Regents exam and an American kid got a 98! I had my share of detention for talking too much, but I was also privileged to learn how to cut and bend sheet metal into a lamp, heat and forge a hook for a barn door lock, make wooden bookends, form plastic disks, and all sorts of practical things which academic schools did not teach. I still have the lab suit I used to wear then in the Chemistry Lab-and it still fits me.

A GRATEFUL REFUGEE KID'S RECOLLECTIONS

Memories from my youth

Between High School and the college years, there were lots of opportunities, not just to learn, but to have fun. At Brooklyn Tech there was time for singing in the All City Chorus and to learn songs like "Good King Wenceslas" and carefully mouthing words like "*Christ our Lord.*" Of course, I had it easy in the chorus, unlike my genius friend, Arthur Wasserman, who was picked to sing some of the tenor solos at Christmas time.

Art was one of those people, who was not only brilliant, but handsome, a star athlete, popular with the girls, and, last I heard, Dean of a University. All this, after winning a scholarship to MIT to study music and the organ, where he got his doctorate in Engineering. Oh yes, we also had the first black G.O. president (the student Government Organization) in, so far as I know, any school in the country. Brooklyn Tech was considered such a tough school that my B-average was considered as an A-when they were considering me for admission to CCNY.

In my late teens I became a counselor at Camp Pine Cone in the Catskills, near Hunter, NY. Mrs. Schwarzschild ran the place in true German-Jewish fashion, with help from her son, Martin. I still remember her camp song: *"Hast Du Tsores, geh nach Hunter. Hunter ist ein wohl Bedruck. Da wirst Du gesund und munter, wenn Du zum Camp Pine Cone kommst."* (or something close to that.) I recall also that when I got to camp, after taking the bus through Monticello and Liberty to Parksville, that I was supposed to be the waterfront counselor.

However, just a few days before, the guy who was supposed to teach horseback riding had decided to go somewhere else, so I was elected to teach riding.

The camp had acquired an old, but willing, nag, and I learned really fast how to saddle, bridle, groom and ride a horse, and teach youngsters, not just how to mount and ride a horse around the ring without falling off, but how to ride bareback and keep the horse under control. I'll never forget the time when one of our wilder boys was riding around the ring bareback, and all of a sudden, the horse tripped. But wild as young Pete was, he gripped with his legs and stayed on as the horse recovered its gait. Whew! I even managed to ride our horse into town (bareback) to get the mail. Of course, there was the time when I drove the camp station wagon into town on a rainy day, lost control around a curve, and flipped the wagon on its side. Happily, there was very little damage to the car or to me (except a bruised ego.)

Once at CCNY, I was happy to join the riding club and go riding with a bunch of other novices in Central Park. That meant that we had to go to a stable on East 62nd Street between 2nd and 3rd Avenue, that was also used by the New York City mounted police to stable their horses. Once we were all mounted up we had to ride the horses across 3rd, Lexington, Park and 5th Avenue into the park. It was my luck to have a 16-hand, high-strung stallion and to bring up the rear of the group, when *afsaloches*, (just my luck) the light changed as we were crossing 3rd Avenue, the elevated train came by overhead frightening the horse, who danced and pranced on the cobblestone street, as cars started honking at us to get out of their way. After that, setting up an illicit jump along the riding trail in Central Park, finding a suitable long pole for a jumping bar, making sure that everyone had a turn to jump, without incurring the notice and wrath of the NYPD, was childsplay.

Sturdy Sons of City College…

The next logical step was a FREE college. At that time CCNY, the City College of New York gave applicants from Brooklyn Tech an extra 10 points on their entrance applications, and I was a shu-in for the Chemical Engineering program. The daily excitement of taking the "A" train from Washington Avenue in Brooklyn to 145th Street in Manhattan and the trudge up the hill to CCNY, past the stone library built with WPA funds during the Roosevelt era, brought us to this romantic series of stone buildings that provided our free education and shelter for the day.

There was the Christmas party in the Chem Lab, when we drank orange juice spiked with 100% Alcohol; but there were also the classes in Shakespeare, where I was the only engineering student, classes whose lessons have stayed with me over these 55 years, thanks to the "old lady" who recited lines that made the speeches memorable. And there was the class in Music, where again I was the sole engineer, where I learned about Bach, and the A B A, B C B of works like his Suite #2 for Flute & Strings, and Mozart's Harp & Flute concerto, whose strains will always be in my soul.

The Strike at CCNY

There was also the infamous student strike, impelled by some of the radical left-wing student members of the AYD, the American Youths for Democracy, who overpowered the Student Forum, to protest the right wing antics of Professors with Pro-German and Anti-semitic views. Funny thing was that on the first day of the strike, the ONLY paper that gave a straight report of the event, was the New York Times–not the News or the Post, or the Herald, or any of the other papers. And it was only later that I found out that our fencing coach, was also an Olympic champion. He gave me some good lessons on being light on my feet.

Those were also the wonderful days when we could go to hear concerts in the bleachers of Lewisohn Stadium at night, after having worked out on the track during the day. Regrettably, progress demanded that it be torn down.

Give 3 Cheers for Old Iowa State, A.M.E.S. I O Way…

My next venue was Iowa State College (now Iowa State University) in Ames, Iowa–an oft-used clue in crossword puzzles. You will probably ask: "what is a Jewish boy from Brooklyn, an engineering graduate, doing, going all the way out to Iowa?" It's really quite simple.

In 1950, when I received my BChE degree from CCNY, jobs were scarce-even for engineers, but especially for Jews. In those days there were no Jews at GE, Westinghouse, or GM. So, a friendly professor managed to get me a fellowship at the Ames Atomic Research Facility at ISC with Dr. Spedding and to study for a Master's degree in Chemical & Industrial Engineering.

Those were the days when people with Jewish names or looks were still refused at some of the more elite Catskill resorts and even at the Elbow Beach Surf Club in Hamilton, Bermuda.

My year-and-a-half at Friley Hall dormitory at ISC turned out to be not just an education, but a wonderful treat. The Chemical and Industrial Engineering classes were fine, as were the cultural opportunities, like hearing and seeing Gregor Piatigorsky from a first row seat in the gym, when he came out to play in the "cultural wasteland" of the middle West. There was also the chance to hear Big Bill Broonzey, who was down on his luck at the time and working as a janitor at Friley Hall, play his guitar and sing in his smoky voice. There was the opportunity to walk across campus, sometimes in 20 below temps, to visit the horses and their newborn colts at the ISC stables–after all, this was an Aggie college. And I was a frequent enough visitor to the stalls so that the mares allowed me into the stalls to handle their newborn offspring.

Then, too, working at the Ames Lab meant working on exciting technology, like separating Hafnium from Zirconium, to create relatively pure product for use in atomic weapons. Of course, accidents happen, and one night, after returning from a thrilling performance by a Soviet ballet troupe, and having to turn down an invitation from one of the ballerinas in the chorus to spend the night with her, as I was on duty to mind the separation column that night, the column overheated and cracked, spilling acid all over the floor. It took a bit of doing to get that process under control and prevent egregious damage–but, we managed. So I kept good records of the separation of Hafnium and Zirconium, and found time to take part in a play, dressed in a Roman toga.

My religious experience at Iowa State

Every Sunday I made time to attend services and hear the sermons at a different church–Baptist, Catholic, Evangelical, Lutheran–you name it. And that was enlightening, even more than the occasional Friday night service held by the Hillel group. There were probably about 1000 Jews on campus, but only about 20 admitted to their faith. Actually, my friend and guide, Reverend LeMoine, knew more Hebrew than any of us Jew kids on campus.

And there was Veisha–the annual celebration, with floats and parades, where I helped shovel mud onto a flatbed truck to create "Mt. Sinai" on which I was to stand, (never knew mud could be so heavy) for the Interfaith float. Every float seemed to be decorated with live flowers–but neither that nor the beauty queen adorning our float made it a hit.

There was the carrying of the torch from Ames to Des Moines, in which I managed to trip during my stretch and crack the torch (ah, the ignominy), and there were the violin lessons from a teacher who didn't think I had much talent, and even after practicing assiduously one weekend, on the promise that I could try vibrato on Monday–she felt I had to keep on practicing–and I gave it all up. Tant pis.

I did, however, learn to back up the flatbed truck, the Interfaith Council had borrowed from a local farmer, for its float of Moses holding the Tablets with the 10 Commandments (standing on a "mountain" of muddy soil, which was much harder to shovel than any of us had anticipated.) Anyway, I did learn to steer the cab wheels one way to get the back to move the other way.

Part 3
WORK

Learning to make Cortisone at Merck's Rahway, NJ plant

Graduating with a Masters in Chemical and Industrial Engineering gave me a chance to get a job at Merck and Co. in Rahway, NJ. There, as an engineering trainee, I did shift work in the factory where tons of ox bile were taken through 39 steps to create minute amounts of Cortisone. Even with 97 to 99% efficiency in each of the steps-0.99 times 0.99, multiplied often enough yields pretty small output. It was an interesting job, rotating through the lab, design, engineering and production, but after a year my boss decided I wasn't cut out for the job.

1952-1955 Helping to build the 280mm Atomic Shell at Picatinny Arsenal

Luckily, at that time in 1952, a job opening was advertised for the Atomic Applications Lab at Picatinny Arsenal in Dover, NJ, not too far away. After driving through a bucolic setting of tree lined roads I reached an old red brick building and was escorted upstairs to an area enclosed by glass windows covered with brown paper and a sign on the door, that said: "Knock, Do Not Enter."

In short order I was ushered in for an interview with the formidable atomic weapons expert (and, I was later to learn, inveterate jokester,)

Bob Schwartz. Half brusque, half smiling, he asked me to define *"Entropy"*–which I managed to do to his satisfaction–and that was the extent of the interview. (FYI-Entropy is the energy in a thermodynamic system unavailable to do useful work.) I was hired to work on the development of the 280 mm Atomic Shell.

Everyone knew about the Atomic Bomb–but a shell with atomic capability–that was a well-kept secret, and a major challenge. My challenge, under the compassionate guidance and tutelage of Milt Epton, our Branch Chief, and Charley Gensheimer, our electronic wizard, was to come up with a "dashpot" design that would fire the Uranium core into the Uranium rings to create the critical mass needed for an atomic explosion.

Not having computers to calculate the plate thickness needed to keep the propellant case from rupturing, it was more or less a question of cut and try, till we got something that worked reliably. After that we needed some more data from the atomic bomb design, kept at the Sandia Labs outside Albuquerque, New Mexico in the Sandia Mountains. So Epton, Gensheimer and I flew to ABQ with our special "Q" clearance and selected "need-to-know" clearance and after much checking and verifying, were ushered into a metal shed housing the super secret drawings. But there was one critical caveat–we could take no notes. It was a case of trying, between the three of us, to remember the critical data that would permit us to complete the design of the Atomic Shell so that it was safe to fire, and effective upon impact.

Well, it wasn't all hardship, because at night we were allowed to go into town, where, when we paid for food and drinks, we got change in Silver Dollars. That was probably about the last time and place where this still happened, and our Silver Dollars became nice mementos.

Meantime, I found time in the evenings to pursue studies leading to a Doctorate in Engineering degree at Columbia University, passing my prelims in German and French, and the orals for my thesis–until my local draft board decided that my 4A Selective Service deferment from working in the Atomic Applications Lab, now that the design was finished, should be rescinded and I was reclassified 1A.

By this time, I was no longer dating the young Branch secretary, but spent most spare evenings with JoAnn Cianello, a nice, slender Italian girl, whose father made wine at home and greeted me in his undershirt when I came to pick up his daughter. I used to take her to local pubs where we would dance the hours away to music from the jukebox—and what a fabulous dancer she was. Quite memorable!

I also had the chance to see the play "Tea House of the August Moon" with Marlon Brando in the role of Sakini, whose memorable line for me, was: "Pain makes man think; thought makes man wise; wisdom makes life endurable."

Putting the "security thing" in perspective—one beautiful fall day in 1953, I drove into the Arsenal through the trees splendidly arrayed in their colorful fall foliage, and decided to capture the scene on my camera. A little later at work, there was a knock on the door and two security guys asked if I had a camera in my car? Sure, I told them, and at their request, went with them to get it out of the glove box. They thanked me and took the camera with them. Next day they brought the camera back with a set of prints, and told me that, though I had breeched security, they would not press charges. Try doing that kind of thing now! Of course, back then people were a little smarter and put things in perspective, and, yes, they listened!

Meeting my Bride-to-be

But then came that momentous Lincoln's birthday weekend in 1955, when I was visiting with my folks at their apartment in Forest Hills, and had a date with Helen Urish, a girl I had met through a friend. She suggested that I bring along another guy so that we could double date with her girl friend. So I shanghaied my neighbor to come along. We met the girls and they suggested a small nightclub on Ocean Parkway in Brooklyn, where we could have a drink and dance.

Little did I realize what a momentous evening this was to be. After the first obligatory dance with Helen, (*obligatory because it was "the right thing to do," not because she was so hard to look at*) I asked this gorgeous little friend of hers, dressed in a daring and stunning red, Ceil Chapman dress with flared shoulders and rather deep décolletage, to dance with me. Oh how we danced and, after a few minutes of rapture, I told her: "You know–I'm going to marry you." (crazy–no?)

Well, she was a little nonplussed by such a dramatic overture, but didn't reject me out of hand.

The evening ended on cloud nine, with our making plans to go skiing together the following Washington's Birthday weekend. I had planned to ski anyway, but with a different group, the Ski Club of America, so it was just a matter of changing the reservations. And ski we did, and danced the evenings away, and I invited her for dinner on March 5th at the Officers' Club at the Brooklyn Navy Yard, compliments of my friend, Johnny Yetto, later to be our best man. Between courses, I got onto bended knee and proposed to her–and, wonder of wonders, she accepted. So on March 5th 1955, three weeks after we met, we got engaged, diamond ring and all.

I took her horseback riding (which she wasn't too keen on–not even to this day) and dancing, while her wonderful parents scrambled to arrange for a wedding only three months hence, on June 26th-a wedding reception still talked about by cousins who were there, and where, by the way, everyone else ate chicken, and this spoiled brat was served his favorite dish of meatballs.

In the meantime my local draft board decided that I should report for duty at Fort Dix, NJ on June 27th. Luckily and happily, after they found out that I was to be married the day before, they granted me a one-month extension till July 27th.

The Honeymoon in Canada

Our honeymoon was spent driving in my old Studebaker to Moodus, CT for our first night and then North to Canada to Lac Archambeau for a glorious week of fun, sailing, hiking and exploring. Then it was South to Montreal and Quebec City where we stayed at the famous Chateau Frontenac and toured this magnificent old town, walking through the narrow streets of the old quarter. And finally back home with a package of Fromage Trappiste which was so ripe that we had to tie it to the back bumper, giving the customs agents at the border a good laugh.

Basic Training at Ft Dix, NJ

When I reported to Ft. Dix for my basic training, remarkably, the officer who directed where we would be assigned for the next two years, was an old friend from City College. He asked if I wanted to go to OCS and spend my next two years in France or Germany, but, knowing that I would thus be spending most of my time with the troops in the field, I demurred and requested Los Alamos or Sandia Base, settling for an assignment at Frankford Arsenal in Philadelphia, where I could not only be living off-base with my new wife, but also be close enough to New York to visit with both parents or have them visit us.

Those ten weeks went by remarkably fast, spending most weekends with my visiting wife at a local motel, learning, as the oldest man in the platoon, to be the guide-on carrier, and even doing the thirteen mile march with a full pack, to come through standing up. As the shortest (and oldest) guy in the platoon, I also learned, perforce, to stretch my step, so that, as guide-on carrier, I wouldn't get my heel stepped on by the guy behind me.

The most memorable thing I learned from Sergeant Mathis, our irascible training sergeant, whose wife, mistress and bank note were all over-due at the same time, was that "a short pencil is better than a long memory" and to "go when you can, not when you have to." Basic and crude, but words that I have always remembered.

Like basic training, my two years at Frankford Arsenal passed remarkably easily, since, to all effects and purposes, I was a project engineer in enlisted man's clothing. In fact when the two years were up, the Arsenal considered my time in uniform as those of a GS-11 engineer and offered me a civilian job as a GS-12. So even though I was just an SP-4 (that's like a corporal), I was the project engineer for the Little John missile fuze. Needless to say there were some hard feelings among the civilian GS-11s in the Small Arms Section (like A. Victor Nardi and the rotund Matthew J. Olevich, who felt that they should gotten the promotion, and who never let me forget it.

Actually, the guy who could really put you in your place (and did so later in my assignment as the project leader for the 5.56 mm Tracer Bullet) was the shop foreman, who threw me out of his shop when I made a suggestion about modifying a batch of tracer mix, as he spat into the batch to make sure it had the right consistency and moisture content. It took a while, but ultimately he did listen and we turned out some great tracer ammo, which was then successfully tested in a new AR-15 (later M16) rifle at Springfield Armory, in Massachusetts.

"Little John" fuze Project Engineer in Uniform at Frankford Arsenal in Philadelphia 1955

There were lots of trips to the field for testing the fuze on the Little John missile, a high priority project, including, on my birthday, when my good wife managed to reach me on a field telephone and patch me through to my parents, from her job as a receptionist at Strick Trailers. Some of those field trips involved visits to Washington to the old DOFL or Diamond Ordnance Fuze Laboratory, later to be renamed HDL–the Harry Diamond Labs. It was at DOFL/HDL that I met my (now) old friend, David Bettwy, the project engineer for the Safety & Arming Device with which our fuze had to be compatible.

Working at Frankford Arsenal from 1955 to 1957 gave me a chance to learn about computers when they were still in their infancy. It also taught me a valuable lesson–to whit: when being interviewed after returning to civilian life, for a chance to receive a scholarship from the Philadelphia Federal Personnel Council for study leading to an MBA at Temple University, I walked in with the other 8 candidates, took the chair at the head of the table, and proceeded to lead them in finding a solution to the problem given to us to solve as a team activity. It must have worked because I got the scholarship and got my MBA two years later.

Writing my thesis on: *"An Attitude Survey of Some Environmental Factors Affecting Productivity of Research and Development Personnel at a Government Laboratory"* almost ended in disaster. (The thesis almost didn't get approved since it included surveys of the staff with questions about the quality of supervision, competence of supervisors, support and encouragement given to improve oneself, extent to which training and skills were being used, and quality of communications in their section, branch, etc. Revealing results, but not always welcome.)

In December of 1957 we moved from a small one bedroom apartment at 6628 North 8th Street in Philadelphia to a brand new Levitt-built home in Levittown, PA, half way to Trenton, NJ, though our two years in the garden apartment in the Oak Lane section of North Philadelphia were happy times. We had adopted a stray dog my secretary had rescued, and "Lady" provided us with many happy diversions, until—. Until one fateful day Lady got out when my pregnant wife was answering the door, and ran off.

So here was this little lady, in a housecoat with wet hair streaming and a big belly, chasing the disobedient pup up North 8th Street, up to the train station, with Lady evading her at every turn, until a kindly gas station attendant at the corner managed to snag the dog for her. Ah, the mixed pleasures of dog ownership.

But this was not to be her last escapade. A few months later, unchastened, Lady slipped her collar and ran off again, evading pursuit. She was gone! Oddly, about 3 months later we were driving through a neighborhood many miles from home, after visiting friends.

We both spotted what seemingly looked like Lady on the other side of the street. We stopped, got out of the car and called to her and she trotted over to us–like "what took you so long?" She had obviously been cared for in the interim, but she never ran away again–well except for the time when we moved to Harbor Road in Levittown and she had come into heat. She had obviously decided to stroll through our new Highland Park neighborhood and came home with a self-satisfied smirk on her muzzle.

Sure enough, on the day in January 1959, when we brought home our second son, Edwin from Lower Bucks County Hospital, she would not yield center stage to Mama Baer, and proceeded to deliver 5 beautiful pups in the middle of the light gray living room carpet–a stain that remained till we moved.

Our first home in Levittown, PA

#1 son, Jeffrey, had preceded Edwin by 18 months, when after 25 hours of painful, seemingly interminable and hazardous labor, our amateur military doctor, was persuaded to do a C-section (Caesarian) to bring forth a very blue baby–just in the nick of time.

Actually, moving into our Levitt built home, beside being our very first house, was quite a momentous series of events as well. After walking through the comfortably and attractively furnished model home off Route 1 and 413, we quickly agreed that at $ 15,800 this 3-bedroom house on a 1/8 acre landscaped plot was a very worthwhile acquisition for our growing family.

When we moved in they were still pouring sidewalks as we tracked mud into our new home. Day by day we watched as crews put down sod; small backhoes would go from backyard to backyard digging holes in precisely designated places. These were followed by crews who dropped a 4 foot willow tree in the middle of the backyard. These same willows, which grow fast, send their roots down wide and deep, and in 10 to 15 years attack the plastic water pipes at vulnerable joints, and sooner or later clog the pipes, slowing water flow to a trickle–but who knew?

However, Bill Levitt, who had pioneered the concept of partially pre-fabbed housing in Long Island for returning World War II veterans, was also a thoughtful gardener. After the willows, came crews who planted fledgling peach and apple trees into other pre-positioned holes in the ground and then dutifully filled in the holes. In the same way they inserted a few shrubs out front, after paving the muddy driveway to the covered garage. Yes–all the comforts of home.

All the Levitt homes were built on concrete slabs into which hot water pipes had been pre-emplaced to provide a warm floor with radiant heat, suitable for children and adults to play comfortably–even with a modest grade of wall-to-wall carpeting in all the main floor rooms. And, of course, he provided window air conditioners. As I said– modest, but very utilitarian and totally livable.

Rather than using the corner room downstairs for "grandma's bedroom," or a room suitable for a child or adult with walking problems, who couldn't manage the stairs, we used it for our study, leaving the adjacent room as a guest bedroom.

That left us with two good-sized dormer bedrooms and a bath upstairs. What more could a young family ask for. And we finally had a place big enough for the parents to be able to visit and stay over–and what lovely pictures we have to jog our memory of those early years.

Dick and Sandy Ludwig lived next door with their 3 daughters, and Sandy would occasionally come over and pull Jeffrey's curly hair, because her three girls all had straight hair. Our new Shul designated me their "weeder leader" and we soon formed a car-pool for the 30 minute trip to Frankford Arsenal.

The only problem with that was that our loud-mouthed, obnoxious neighbor and his equally mouthy wife, sometimes made the trip miserable for all of us. Cohen would smoke his cigar in the car, making trips, especially in the winter, a real trial. Between them and my curmudgeon boss, Edgar Beugless, who showed me drawings he had signed the year I was born! I soon developed a nasty skin rash.

We went to see the lead dermatologist at University of Pennsylvania and he just asked questions and listened to my tales of woe before prescribing Dexidrin and Trilafon to be taken as needed. Happily, they worked, and after leaving Frankford Arsenal I put the pills on the shelf and never needed them again.

In 1962 my parents, now comfortable retired, moved from their Forest Hills apartment to a neat little home in Levittown to be near their grandchildren. Unhappily, my father's emphysema and lung cancer, the price of years of smoking, exposure at Buchenwald and working in a smelting plant, soon debilitated him, and though treated with great compassion at Philadelphia General and Lower Bucks County Hospital, he succumbed in September 1963, just a few months short of his 70th birthday.

However, both sets of parents happily celebrated the arrival of our little girl. When Dr. Morty Nelson told Mama Baer that she had a girl, Shirley told him that he could now tie the tubes–no more children. One small problem, this beautiful little girl had two very noticeable hernias, which were quickly and neatly patched by, to us, an unknown pediatrician, Dr. C. Everett Koop. But more of that later.

Chief High Explosives and Pyrotechnics Small Arms Ammunition Section 1957 to 1960

As I noted previously, there were some hard feelings among the GS-11's who had been passed over for promotion, when I got the GS-12 slot that they felt they deserved, but fortunately we had a wonderful Branch chief, Alfred S. Hitner, who had made the decision to hire me and who provided the friendly guidance to keep me going. Good thing too, because my Section Chief, Ed Beugless, was not a gentle soul. He took delight in showing me drawings he had signed the year that I was born. Now there's something to put you in your place.

But, as I learned about small caliber ball bullets, I was also introduced to tracer ammo, grenade launching ammo and even blank ammo–both metal and plastic. We had to compare the Norwegian *Bakelittfabriken Lospatroner*, the French *cartouches á blanc, and* the German D.A.G. *Platzpatrone* against our American version-see photo on page *102.*

An interesting aspect of this work, was the ability to work with engineers at Springfield Armory, in Massachusetts, on the design and test of the blank firing attachment. It was an interesting professional association that was to continue through small caliber weapons development over the years, as I moved to LWL, SASA. AMSAA and HEL, to ultimately providing overview of their R&D projects when, years later, I was put in charge of Manufacturing Methods and Technology at the Army Materiel Command in Virginia.

Part 4
Life in Maryland

The 1960 Move to Harford County, Maryland

Shortly after Susan was born, I was offered a GS-13 job at the newly formed Limited War Laboratory at Aberdeen Proving Ground–sight unseen! It was heart-rending to move so far away after my folks had severed their ties to friends and family in the NY-NJ area to be near us, but the opportunity was too good to turn down. And so we moved in February 1963 to a rental home in Bel Air, MD and soon bought a lot on a good sledding hill near the school and shopping areas.

Here we faced a new experience–building a new home from scratch and having to make decisions about room sizes, appliances, cathedral ceilings, window styles, brick colors, appliances, the location of a well, a septic tank downhill from the house, and oh so much more. Our builder, David Joesting, fortunately had a talented wife, who took a plan we liked, flipped it over and David started digging for the footings on our ¼ acre, heavily wooded lot. And thus began what Grandma Fussia's sister, Tjotja Fanya, was later to call our Daicha in Harford County, MD.

Though Susan was still a baby when we moved in, Jeff was now a sagacious 6 and Edwin a dynamic 4 year old terror. They loved the winters when they could sled down the driveway in the snow, and when we could block the snowplows, all the way down Stuart Terrace to General C.D.Y. Ostrom's house and the pond. And then there was the day one spring when the cows broke through the fence across the road and wound up tromping up our lawn, searching for greener pastures.

It was a time when I made many trips in our old Cadillac convertible to scrounge used bricks from a demolished church near-by, to build a patio outside our dining room and another down below, off the basement family room. Of course, I was very careful to make sure that they were perfectly level–only to learn that was not smart, and I had to take up all the bricks, and smooth the sand base so it would slope away from the house, and then lay the bricks down all over again. That was a learning experience.

The Harford Jewish Center

At that time the Harford Jewish Center was located in an old American Legion Hall in Perryman near Aberdeen on a narrow country road that dead-ended at the Chesapeake Bay. Makeshift was about as respectable a term as you could use to describe this facility. The congregation soon realized that we would have to build a "real" synagogue, complete with schoolrooms, kitchen and social hall. We were lucky enough to find a hilly piece of property that overlooked a good rural road and was less than a mile from the Havre de Grace exit of Interstate 95.

Don't ask me how, but, as a member of the board, I became the building chairman, and thanks to another good Italian builder, we got off to a good start. Oh yes, there were glitches–like the time I came to the site to check on the location of the footings. When I stood on the high point of our property, where the Shul was to be built, I asked our builder where the footings were, and he pointed–over there!

Good thing I asked–so they had to dig new footings in the right location and within a reasonable time up went the walls of the sanctuary, the social hall, the kitchen, school rooms, and the rabbi's room. The wood paneling and stained glass windows were installed at the East end and the Ark was moved from the old building and tenderly placed in its new location. And this is where our children were subsequently Bar and Bat Mizvoh'd.

Meanwhile I got to be vice president of the congregation, until one fateful night when the congregation met to decide the fate of our Rabbi, who was beloved by many for the way he worked with the children, and who outraged many others by ordering ham and cheese sandwiches at the Deli in town, his profligate ways with money (and his less than professional ways with women.)

Needless to say, the discussions whether to retain or fire him became acrimonious, and by 2 AM, the president resigned in anger, leaving yours truly as the new president. Ultimately, the Rabbi was ousted, and after a couple of temporary Rabbis we were fortunate to find Rabbi Kenneth Block, newly ordained, who served the Shul well for some 25 years.

Funny, the way we found him. He was one of the candidates whom I had the chance to interview at the airport on a stop-over in Cincinnati while I was on a business trip.

We clicked immediately and I hired him on the spot and never regretted the decision. Fortunately, the board didn't give me any grief over that precipitous action–especially after he passed his Shabbas try-out with flying colors.

Life in Harford County, MD

By the time Jeff had reached 13 or 14, he decided that we need to finish the family room and create a bedroom for him downstairs, away from the room he had shared with his irascible, long-haired brother. (Shirley used to chase him around the house vainly to try to trim his long locks.)

By the time Jeff was a Junior in Bel Air High School, he had also become quite the computer expert–to the point that he would get up at 3:30 AM and go to school to run their computer programs. By the time the teachers got in he was able to tell them about their classes, school schedules, etc. Although we had pretty much come to terms with his schedule, it was after we had returned from a trip that Grandma Fussia informed us about Jeff's impossible hours and how worried she was about him. She also notified us in no uncertain terms that we should never leave her to take care of these three wild ones again.

Little did she know that on an earlier occasion, when Edwin was about 10 years old, we had left him with the parents of one of his favorite girl friends, Meg McLaughlin, while we attended a wedding. Upon our return that evening, Meg's mother greeted us at the door: "Now don't worry, Edwin is all right." Then she told us that while Meg and Edwin had been playing with some farmer's children in the farm field, one of the boys had gotten angry with Ed and thrown a mattock at him. Thankfully it missed his eye and did not hit his forehead hard enough to do much more than break the skin. Still and all, it was enough for us to break down once we got him back home and into bed. But that was Edwin–ever the daredevil.

Now you have to remember that Bel Air was south of the Mason-Dixon line. Here we were, the first Jews on the block, and it took my good wife several years to get our, otherwise enlightened neighbor, to stop saying Nigger. This must have been especially hard for her when we opted to invite Carlester Sherman, a young black, inner-city boy from Baltimore, to spend a few weeks with us in the country during summer vacation.

It was still difficult in those days to find a pool where we could take him to swim, other than the one at the YMCA and at the integrated Aberdeen Proving Ground facility. But we persevered, and Carlester came back to visit with us the next summer, when his black face was more readily accepted.

Meanwhile, a recycling program was started by John Brown, the black football coach at Bel Air High School, where people would bring their bottles, cans and newspapers on Saturdays and we volunteers sorted it into bins. *(See page 103, that's me with Colonel Ray Isenson, SASA's CO recycling)(Small Arms Systems Agency)*

It was also a time when some good people started an Interfaith Council and black and white folks used to meet and talk and learn that except for the color of their skin, they had an awful lot in common. Maybe that was why our elderly cleaning lady, Emma Cornick, whom we and the children always called Mrs. Cornick, invited us, as the sole white guests to her wedding reception, when she married a local widower.

The Limited War Laboratory (LWL) (1960-1962)

Of course, there was work too, and a carpool of like-minded folks who used to drive the backroads to Aberdeen Proving Ground (APG) each day. I must add that when it was my turn to drive, I would occasionally drive in the white Cadillac convertible or later a red Ford Mustang convertible–with the top down, summer and, yes, winter. Funny thing was–no one objected, but we all just dressed for the 15-mile drive and everyone considered it a lark. I guess, working at LWL, we were all a bit eccentric–especially some of our crazy helicopter pilots–Major David Hayes and Major Vincent Oddi, who taught me a lot.

Vince was the more cautious pilot, who let me fly the Huey when we were out on a mission, and even let me take the controls of our two-seater plane when we flew down to Fort Belvoir for a meeting. It was a bit rough going, with the warm air rising, causing turbulence over Baltimore, but I learned how to hold the plane steady.

Dave Hayes was another story–a daredevil, who would swoop over our house in Bel Air as we headed North to Picatinny Arsenal or the test range, and who thought nothing of flying our HU-1 over Chesapeake Bay with the doors open and the aircraft laid over 90 degrees on its side! even when we took our two boys for a joy ride–which in those days you could get away with.

Beside our highly educated Commanding Officer, Colonel Bob McEvoy, we also had seasoned field officers like LTC Austin Triplett, Jr., head of the Military Operations Division, who would slash the adjectives out of my reports and taught me the inherent preferability of the KISS principle. He always insisted–"keep it simple, stupid"–and I learned.

We also did some of what, in those days, was considered to be pretty novel–we took secretaries, like Jane Preston, shown in the photo section, out to the field periodically so that they could see first hand what all the letters and reports were about. Sometimes we even used them to model some of the projects, such as when we asked Jane to pose with the Communication and Protective Helmet we had developed to connect with the AN/PRC 77 Radio-see photo on page 104.

The Munitions Branch-Elmer Landis

In the Munitions Branch, part of Development Engineering Division, we labored to make life and combat easier for our troops in Vietnam. That included projects like a "floating smoke grenade." Often emergency smoke markers (to call in aircraft to rescue or attack a position) would land in the water, rather than on dry land. We did this by creating a "ballute" that inflated when the grenade hit the water. We also developed low-noise handguns that could be fired in the tunnels where Viet Cong were hiding, without blasting the eardrums of the shooter. Our efficacy was determined by the fact that an LWL engineer went to VN for a 3-month tour of duty to learn, first hand, what the troops really needed.

We developed barometric and timer activated propaganda leaflet bundle breakers to help our aviators drop leaflet bundles from high altitude, set to open at a lower altitude to assure optimum distribution over the countryside. There was a need for smoke markers needed to mark targets, that would float in swamps or rice paddies, either when dropped from an aircraft or dispensed by hand.

There was a story that went with the bundle breakers. When we went to test them from a cargo plane flying over the bay at Florida's Lackland Air Force Base, I went up in the plane to set the timers and then toss the bundles out over the bay, so we could measure the altitude at which the bundles opened.

Well, after throwing out the first batch, I unhooked the safety harness that hooked me to the opposite wall of the aircraft and went forward to fetch another batch of bundles, with barometric devices to drop on the next pass. Casually, I went back to the door and got ready to push them out on signal, when our observer on the ground saw in his telescope that the black Sergeant supervising me almost turned white when he noticed that I had failed to cinch up my safety harness.

It made for a good story, and we celebrated that night at an unimpressive, local shack, where the owner and part-time school teacher displayed racks of steaks for the guests to pick, and which were served with his selection of 1959 *Piesporter Goldtröpfchen*. That happened to be a five star year for *Mosel-Saar-Ruwer* wines and I proceeded, after returning home, to buy every bottle I could find at the state liquor stores in Harford and contiguous counties–all the way up to York, PA. As to Elmer Landis, our branch chief, he was the one who insisted I use my middle name, Larry, as he already had 2 Johns.

The LWL Munitions Branch (1963-1969)

When our liaisons in Vietnam reported the need for indigenous armor for trucks, we found that the local Ceiba Bombax or banana wood did a great job in stopping bullets, by actually absorbing them. It was thick and a bit awkward, but lacking armor plate, it provided quick protection for the troops and it worked-see photo on page 105.

For the more serious attacks, we found that the Russian developed High Hard armor plate would stop a 7.62mm bullet at half the thickness of standard (MIL-A-12560) Military armor plate and $1/4^{th}$ the cost.

The only problem was that since the Russians had developed it, "it couldn't be any good," and besides it was hard to work with and to weld. But we persevered, and with the backing of our good CO, Colonel Bob McEvoy, we fought the bureaucracy all the way up to the Asst. Secretary of the Army to get approval to use it on our jeeps and the Yabuta junks patrolling the k*longs* or canals.

Ultimately, we managed to save the Army some $12 Million, and I received a medal and a citation in 1995 for "outstanding achievement in the application of new materials to provide vehicular protection against ambushes, delivering lightweight modular armor kits to the field in the unusually short period of nine months." What you see here was a 2 ½ ton truck with armor plates on the sides and door, with vision slots cut out to fire back against the Viet Cong, and 2 ¼" thick, 17 #/sq.ft. glass-plastic laminate that would stop a caliber .30 bullet at 100 yards. The XAR-30 high hardness steel was a rolled, homogeneous steel plate, heat-treated to a Brinell hardness of 480 to 500. We also experimented with "field expedient" armor, such as Teak or Balsa wood faced with locally made 1" thick, clay tiles, that would stop bullets for a while, and some local, indigenous wood, called *fromager* or *Ceiba Bombax*–anything to buy time as our troops ran a gauntlet of enemy fire.

The best part was the $ 1200 reward that went with the medal, for saving the Army's money, which helped to pay for paving our new driveway. No, really the best part was working with a small company in Florida that fashioned high hard armor plates to be used on the boats our troops used on the Vietnamese canals. They learned how to fashion the plates, cut out vision ports and to weld on hinges so the plates could be fastened to the deck and quickly erected when the troops came under fire. And so one day, a young Lieutenant came to my desk with his arm in a sling, and with his good hand, shook mine and said: "Mr. Baer, thank you. I owe you my life. When we came under fire, your armor plates stopped the bullets that would otherwise surely have killed us." That was the best kind of reward.

There were lots of good times, like driving to Picatinny Arsenal in Dover, NJ with my friend Bud Pierne, who was always full of stories and good humor, and who cautioned me that if we were stopped for speeding to "look pregnant." He also kept me from getting a swelled head when I started rattling off about something I didn't really understand–like the Junghans movement in a fuze.

Funny, how 40 years later one remembers the events so long ago–like the work we did for ARPA, the Advanced Research Projects Agency, which had men riding around Vietnam in International Harvester Scout vehicles, to learn what the troops really needed and how our products were helping them.

For them we developed high hard steel door armor plates and glass-plastic laminated windshields and door windows, about 4" thick. They were heavy and awkward–but they worked to stop bullets from getting through. We had to beef up the engines and shock absorbers to carry the extra load, but saving lives was paramount. Then ARPA also asked us to put armor on the bottom of the Scout, in case they ran over a mine, and here we almost added injury to insult.

Fortunately we had learned about a Sergeant, whose jeep had run over a mine and peeled back the bottom armor plate, severing his legs so surgically that he was, in his shocked state, not even aware it had happened. Thus with this tragic lesson learned, we chose to put a high yield strength steel plate under the Scout, which would absorb the blast of most road mines and buckle–but not tear, as the regular armor had done.

Unfortunately, we have seen again in Afghanistan and Iraq, that, as we got smarter in up-armoring our vehicles, the guerillas have learned to increase the lethality of their roadside *improvised explosive devices* or IEDs, to disastrous effect. Of course, beside the Munitions Branch, other branches were busy developing such unusual and exotic devices as a tree-top helicopter landing platform. This platform was carried, slung under the helicopter.

One final project I have to show is the jungle tree-top canopy platform that was designed so that helicopters could land on the 4-tier canopy to discharge troops, who could then work their way down to surprise and attack the enemy.

When the helicopter came to a site where troops were to be deployed on the top of the triple canopy, they lowered the platform, which permitted the troops to descend onto relatively solid footing, and then rapel down through the canopy to the ground.

Other *Munitions Branch* Projects

There were other interesting and novel projects that will never make the papers. For example, when our troops needed a nighttime target or a position marking ground flare to mark a rescue pick-up or a target location, much of the time in Vietnam it would land in a water-covered area. So we developed a 40mm floating flare, wrapped in a *ballute* that served to retard the fall of the round and provide an upright floating surface upon water impact. The flare then burned from a chimney that extended through the center of the *ballute*. And it worked.(see photo of ballute on page 106)

We also developed "less lethal" ammunition that would later be used to quell riots and slow down rioters without wounding them. There were such items as "beanbags" filled with lead shot and rubber bullets that were actually tested by volunteer police officers, who attested to the rounds' efficacy.

Another project that is still in use today, was the development of a "bean bag" which was not as potentially lethal as rubber bullets. Bean bags could be fired from a 40mm grenade launcher and when tested by volunteer policemen (after having been tested against animals) caused them to wince when they were hit on the upper arm from 50 feet. After rubbing away tears, they said that they'd love to have them for use in a potential future riot.

Finally, our troops needed to get up into the trees that dominated the terrain, as shown above, either for observation posts or to set up an ambush. So we developed a "grapnel, with line, propelled" that would launch a grapnel hook with 400 feet of 5/16" diameter nylon line and propel the hook 150 feet up into a tree or up to 220 feet horizontally. Climbing devices that were packed with the grapnels then made ascent into a tree possible.

SASA–the Small Arms Systems Agency (1969-1972)

With the experience of developing small caliber ammunition at Frankford Arsenal and at LWL, it was an almost natural transition to be invited by Colonel Walter Rafert, who headed the newly formed Small Arms Systems Age, to be the Chief of their (6.2 funded) Exploratory Development Division, promoting me from a GS-14 Chemical Engineer to a GS-15 Physical Science Administrator.

Looking very young at 41, I decided to grow a goatee to try to look older, and determined to hire the best, most senior engineers and physicists to staff this new Division, demanding that each had to be qualified in at least two domains, and to be smarter than me.

I was also determined to try to hire a woman physicist, though that fell through when she called on the day she was due to report for work, saying that her husband would not move to the Aberdeen area. Well, I tried. My plan was to make assignments to each of my people and then let them do their thing without interference, periodically reporting their accomplishments or problems to me, if they needed my help. I just didn't want any surprises. I also made sure, after giving a verbal assignment, to ask them: "tell me what you *think* I said." This sometimes resulted in unusual responses, indicating that either I did not express myself effectively or someone wasn't listening well.

Back in those early days, before women's lib, I also had a unique experience in hiring a secretary. Those were the days when you could still demand that they pass a test in taking dictation and typing. So when I was interviewing a candidate for Division secretary, in came this kid from the country, Emily Waddington, with long, unfashionable hair, big glasses and a dress, more suited to the farm than an office, carrying a plastic folder. She was painfully shy, almost ungainly, but yet poised. When I asked for references, she opened the folder, showing in neat, plastic sheaths, examples of her work in school and letters of such high praise and commendation from her teachers and after school jobs as to set me wondering–can this kid be for real?

As I had done with each of the other candidates, I gave her some brief dictation. After taking the message down in shorthand, she read my words back to me, typed them up, neatly, swiftly and absolutely error-free. What's more, she even "dared" to change some of my words–for the better. I hired her on the spot and in the years to come Emily was the only secretary I ever had, whom I allowed to edit and change my words (except for my current "editor-in-chief," my good and verbally skilled, painstakingly careful wife.) (Emily has since been promoted several times and wound up as a GS-13 Administrator–something almost unheard of in those chauvinist days. But then, as I said–she was good!)

A GRATEFUL REFUGEE KID'S RECOLLECTIONS

I guess that our projects were successful enough, that after three years, in 1972, the new SASA CO (Commanding Officer), Colonel Raymond Isenson, asked me to take over the poorly performing (6.4 funded) Advanced Engineering Division as a GS-15 Supervisory General Engineer. It was a nice challenging assignment and I also managed to bring that Division around to be successful.

In fact, it was during this time period that my boss, Leo Ambrosini, our Technical Director of Italian Swiss background, asked me to come with him on a tour of some of Europe's best small arms firms, such as Heckler & Koch, Mauser, Diehl and the Belgian firm of *Fabrique National de l'Armes de Guerre*, known as FN. Since America was the dynamic market for small caliber ammunition and guns, we were accorded the royal treatment at each factory. Since this was my first trip back to Germany, I made sure to wear a large Star of David in my lapel, which invariably caused the reaction of: *"Ach, vosn't that a terrible zing zat happened. But ve knew nozzing about it. Nozzing."* Ah, well, they tried to be nice, showed us all their latest systems, in hopes of making a big sale, treated us to meals at their finest restaurants and hotels.

I remember once when they served us a fine *Danziger Goldwasser* as an *aperitif* before lunch and it burned my throat, so that, when no one was looking, I dumped the contents of the glass into a planter. Poor plant.

An interesting sidelight occurred at FN, when, as the VP was touring us through the plant, one of the foremen asked him, in French, whether he wanted to show us their new mini-mitralleuse or machine gun? And he replied, in French, *"ah non, pas encore."* (not yet) And sneaky me, who had not let on that I understand a bit of French, gently asked: "eh, pourquoi non?" (why not?) He blushed all shades of red and then, after explaining that he thought it was premature, showed us their new weapon, which was actually so good, that the U.S. Army later bought many thousands of them for our troops. That is how history sometimes turns on a few words.

The Demise of SASA & Return to the new LWL (1972-1974)

Unfortunately, someone at the Army Materiel Command decided about this time that they really didn't need SASA, and so, in 1973 it was disestablished and abandoned. Fortunately for me, LWL was glad to take me back at my GS-15 salary and switch me from a Supervisory General Engineer to a Supervisory Chemist, to head the Chemistry Laboratory. So I had come full circle. After all, my first degree from CCNY was as a Bachelor in Chemical Engineering.

But the *Land Warfare Laboratory* had also outlived its usefulness and in 1974 it too was disestablished. Now that was interesting, because in our storage shed we had accumulated all sorts of toxic materials, and innocuous, but noxious materials, like gallon jugs of Nuoc Mam sauce, that we were studying as how to make the Viet Cong who used it, sick enough so that they were unable to fight.

Those chemicals that the local schools could use, we gave to them, and those they couldn't we poured into the ground–try doing that today? The grass will never grow in those spots again.

The "other" Branches

The *Mobility Branch* developed such items as a lightweight, High Mobility Wheeled Vehicle, sporting 4-wheel drive, that could be used for high speed, recon or security patrol duty, and the tree top helicopter platform (see page 51.) Other mundane items included a scratch-resistant plastic window for trucks that had to maneuver through the jungle.

The *Environment & Survival Branch*, besides developing lightweight shelters, developed an "Automatic Personnel Verifier" to provide rapid verification of people requiring access to secure areas. These were basically fingerprint readers (now in common use) that stored optically encoded and scrambled fingerprints, to unlock doors to secure areas.

In the *Applied Chemistry Branch*, which I was leading at the time, we developed simple colorimetric tests to give a positive identification of hashish, Heroin or marijuana users. More relevant to today's dangers, we were asked to develop a simple explosives detector to locate small factories that were reclaiming explosives from dud munitions. The aircraft carried "ion mobility spectrometer" offered one of the most sensitive and specific techniques for detecting airborne effluvia of explosives. That would seem to be a useful device to detect bomb and IED factories in Iraq.

One project that only made it through the feasibility study was a "Snow Stabilization Technique" to help helicopters landing on snow from being blinded by the "white-out" from blowing snow. This was requested by one of our liasison officers in Alaska. We actually found that methanol and water surfactants effectively congealed fresh snow and increased surface bearing capacity from 7 lbs./sq.ft. to 130 lbs./sq.ft. and resisted air blasts from helicopter downwash up to 60 mph.

The *Biological Sciences Branch*, trained dogs to detect explosives and narcotics, and delivered them to the Provost Marshall at Ft. Benning, Ga. in September 1972. They also trained dogs to locate and recover bodies in military or civil disasters.

I've noted only some of the more abstruse projects that we worked on in this very unusual facility. Happily, as in the Human Engineering Lab, we were able to respond quickly to field requests, with the result that each day brought new challenges, and an atmosphere that was a delight to work in. It was this kind of collegiate atmosphere that I tried to implement in all my assignments as a civil servant. Setting challenges that made life interesting for the staff and giving credit, publicly, when they succeeded (chiding privately when they did not) gave me and my staff at every facility a feeling of wanting to come to work every morning.

AMSAA & The Human Engineering Lab (HEL) (1974-1976)

Well, again, my luck held out and the irascible, but charming director of AMSAA, the *Advanced Materials Systems Analysis*

Activity, Joe Sperrazza, agreed to bring me over, once more as a GS-15 General Engineer, so that I suffered no loss in pay–at least for two years.

Once more good fortune was to be at my side, when Joe told me that in a few months he wanted me to visit some of our military bases in Germany, where, of course, I knew the local language, and Italy, where I knew not a word, to see how well our troops were being served by the new materiel we were sending them, and to ascertain their problems.

Having some two months to prepare, I started immediately, to listen to audio tapes to learn Italian. Luckily, it took and by the time we got over to Milano I was able to ask: *"dove e il gabinetto?"* (where is the bathroom?) and other important stuff like *Buon Giorno and Grazie Tante and Prego.* This would serve me very well in later years (1984) when I was invited to assist Oto Melara in La Spezia to prepare a proposal for maintaining tanks the U.S. Army was using in Europe.

I bring up HEL because the concept of human engineering, later called human factors engineering, was still quite novel in those days, but under their distinguished director, Ed Weis, they studied the interaction between weapons systems and the humans who had to use them. Presumably, this work led in later years to OSHA and their investigation of human interaction with all sorts of systems and the effect these systems had upon the humans. Happily, just as my two years as a GS-15, serving in a GS-13 slot were about to come to an end, we received a call from AMC headquarters telling me: "You're Hired."

At The Army Materiel Command HQ 1975–1983

To be hired, sight unseen, for a position at any place, let alone the Army Materiel Command, seemed so unusual to both my wife and me that she insisted in coming with me to the personnel office when I reported for work in Alexandria, VA. Sure enough, though, they informed us that there was a GS-15 vacancy in the Manufacturing Technology Office, and that my qualifications matched their needs so exactly that my name rose to the top of the heap.

Since our office was on the 9th floor, this gave me a good opportunity to get my exercise every morning by climbing the nine flights of stairs, rather than taking the elevator. After the retirement of the Colonel who ran the office when I came on board, a man who drank his lunch every day, so that you did not want to discuss problems with him in the afternoon, I got a chance to be the acting chief.

The fun of this assignment in Manufacturing Methods & Technology (MM&T) was that we controlled an annual budget of about $ 100 million a year and had a chance to review proposals and project reports for all the major commands–aviation, electronics, munitions, vehicles, including tanks and trucks, and including esoteric projects aimed at protecting the troops at the Natick Research Labs and Watertown Arsenal.

Most of these proposals were well thought out–they had to be if the proponent hoped to get funding support. And the quarterly reports challenged the reviewer in our office to remember what he had read about the particular project in the past.

Funny, how some people think they can get away with putting a new date on an old report. But that is exactly what one smart-alec did, and wound up with a severe reprimand–it's always been tough to fire a civil servant except for some really egregious or even criminal behavior. But the real fun was visiting the labs and hearing the oral reports on each project, seeing the physical results of their work, and discussing how they planned to ultimately achieve their goals.

Granted, some folks were scared to present their work because a) I had read their reports, and b) knew enough to ask probing questions. However, most of the engineers relished this opportunity and were happy to show off their results, or even to ask for advice when they ran into a problem that seemed to confound them or came outside of their purview.

The "other" Commands—ARMCOM, AVSCOM, ERADCOM, MICOM, TACOM & the Arsenals and MTAG (Manufacturing Technology Advisory Group)

Conferences

By the time I retired from AMC, the budget for ARMCOM, located at Rock Island, IL, along the Rock River, had shrunk to $18 million for Munitions R&D at Picatinny Arsenal in Dover, NJ and $16 million for Weapons and support R&D at Rock Island and the various depots.

AVSCOM, the Aviation Systems Command in St. Louis, MO did a lot of good work with just $10 million. During one of my project reviews out there, I was delighted to find that they had perfected their aircraft assembly technique to the point, where the laser alignment was so perfect that no adjustments were needed to the aircraft body. They had basically adopted the Japanese dictum of "building quality in" rather than "weeding bad parts out and adjusting improper assemblies" on a sampling basis.

Next in MM&T budget magnitude was ERADCOM, the Electronics R&D Command at Ft. Monmouth, NJ, whose $9 million provided such 1983 vintage avant garde items as *"long life emitters for fiber optics," "low cost silicon photo-detector modules"* and *"YAG Laser Rod Fabrication."* 24 years later we read:

"Optical pumping of Nd:YAG lasers is of particular interest because they have become widely accepted for industrial and medical use, along with the CO_2 laser. The laser active material which, in the case of the Nd:YAG laser, consists of Neodymium ions accommodated in a transparent YAG host crystal (Yttrium Aluminium Garnet). Where only up to few years ago, Nd:YAG lasers were mainly excited using powerful discharge lamps, optical pumping with laser diodes is becoming more and more important. This is because powerful laser diodes are nowadays available economically and they emit light at high optical power levels with a narrow spectral bandwidth, which matches perfectly with the energy levels of the

Nd:YAG crystal. The great advantage over the discharge lamp is that the emission of the laser diodes are almost completely absorbed by the Nd:YAG, whereas the relatively broad spectral emission of discharge lamps is absorbed to only a small extent.

(*source: Photonics for Engineers from SMC's* Mechanical Electronic and Optical Systems) (I have included this 2007 description just to put in perspective, the advanced nature of some of the work the MM&T program was funding.)

The bulk of MICOM's (Missile Command) $ 5 million was devoted to such projects as *"Real Time Ultrasonic Imaging"* of the machining process, as was common in Japan, and *"Computerized Production Process Planning and Execution for Machined Cylindrical Part,,"* in common use in German factories such as MBB (Messerschmidt-Bölkow-Blohm.)

AMMRC, the Army's Materials & Mechanics Research Center, used its $ 4 million for such tasks as *"Transportation Vibration Testing"* (a major challenge for the off-road transport of electronics in Korea, where low frequency, high amplitude vibrations during truck transport would result in rendering a component useless by the time it got to its destination.) (as opposed to the high frequency, low amplitude vibration encountered in air shipment.)

They were also involved in *"Infrared Testing of Printed Circuit Boards"* and *"Finding Quench Cracks after Heat Treatment."* They also did early studies on cooling uniforms for personnel who had to work in very high heat surroundings–something equally applicable to our space program.

And among lesser-known facilities, such as Watervliet Arsenal in upstate New York, there was a demonstration of how technology often doesn't change all that much. The engraving of gun barrels in 1979 was exactly the same as it had been in 1896-see page 106. But even then, successful efforts were made to reduce energy consumption in the forging and heat treatment of large caliber gun barrels by as much as 50%, using energy recovery for the furnaces and heat treat line. And there were many pleasant memories, such as taking after-lunch walks along the Rock River, which bordered the Rock Island on the North, with my friend Arnie Madsen, hands behind the back, in European style. And there was the time when I stopped in to see my friend, MG Ben Lewis, the ARMCOM commander. As I walked in to his office, he croaked out "Don't touch me!" When I asked him "Ben, what's the matter?" he said "I've got a terrible cold and I don't want you to catch it." Now there's a friend. *(In the center section you can see how gun barrel numbers were imprinted 100 years ago and in today's age–not much improvement there.)*

Recycling Toxic Wastes

One of my very favorite projects concerned the toxic wash water from the TNT lines at the *Radford Army Ammunition Plant* that was being dumped into the Radford River–to the distress and anger of the folks downstream in Radford, VA, who had to clean up the water before they could drink it.

The Munitions Command submitted a $10 million proposal to set up a filtration plant for all the wash water, but someone else came up with the bright idea of recycling the wash water, with the result that for under one million dollars, Radford was able to set up a system which collected the wash water from all the TNT lines and strip out the 1000 pounds of nitric acid a day per TNT line.

They were able to re-use this acid in the production process, and, to the delight of the citizens of Radford, downstream, return excess water, properly cleansed to the Radford River. And the project paid for itself within six months from the recovered acid.

Educating the Government Manager-1978

One of the benefits of being a Government "executive" is the periodic training that is provided. For example, in August of 1978, we were invited to a *"Symposium on Government & The Economy."* Before you stifle a yawn, consider some of the speakers.

A GRATEFUL REFUGEE KID'S RECOLLECTIONS

Our first speaker was a very dynamic Professor from the Georgetown University School of Business, by the name of John Sauter, who gave us an *"Overview of Economic Issues."* If you check his credentials they will show you:

John Sauter specializes in presenting workshops on economics and cost-benefit analysis and has done so for a variety of organizations. His career spans more than thirty years as an educator and researcher, with extensive involvement in designing, presenting, evaluating, and managing courses in how to conduct a cost-benefit analysis.

I couldn't imagine a better person to provide a dynamic presentation on what could be a very dull subject for an engineer. He was followed by "The Honorable" Lyle Gramley, who, at the time, served as a member of the White House Council of Economic Advisers. But the best was yet to come.

After lunch we were treated to a report on *"Congress and the Economy"* by none other than the indomitable Alice Rivlin, who was then Director of the Congressional Budget Office. She was a fantastic speaker then, and these days, as a Director in Economic Studies at the Brookings Institution, and still an expert in Fiscal and monetary policy, social policy, urban issues, still enlightens the public on the PBS' "Nightly Business Report."

A hard act to follow, she was nevertheless matched by Leon Keyserling, the former chairman of the Council of Economic Advisers. From 1946-1953, he had served as chairman of President Truman's Council of Economic Advisors. "During this time, he helped formulate economic policies which enabled the nation to smoothly adjust from a wartime economy to a peacetime one while meeting society's demands for increased housing and schooling without inflation."

As if this wasn't enough to overwhelm and delight–the next afternoon opened with a brilliant talk by Henry Wallich, a member of the Federal Reserve System Board of Governors from Boston, and a fantastic talk on the "Future of the Economy" by Professor Robert Lekachman, from City University of New York's Lehman College. But the best was yet to come.

The symposium concluded with an Invitation to attend a Special

Presentation on "Economic Issues and Regulations" by the extraordinary, dynamic Alfred E. Kahn, who, at the time, was Chairman of the Civil Aeronautics Board and later, advisor to President Carter on Deregulation. He was then, and still is, a dynamic speaker, who can always be counted on to thrill an audience with his stories, and to rattle off facts that you wish you could remember. He left us with the challenge of *"Beneficat Emptor"*–that we were to be dedicated to the Public Benefit and Convenience–not a bad challenge.

Another *Symposium on Government & The Economy-1980*

This symposium in February 1980, addressed the same area as the one two years earlier, but with a far different approach. The speakers addressed such novel topics (for that time) as "the aspirations of black and women workers vis-à-vis white men" and such concepts as "upward mobility," "equal pay for equal work," "making work interesting and satisfying," "the limited capabilities of 'uneducated' college graduates," "blacks views of other blacks," and "the unwillingness of some people to work." The panels also noted that "productivity is the manager's, not the worker's responsibility (per Peter Drucker) and such topics as "second careers for women at 40, who can't compete with fresh 24-year olds" and "the need for useful challenges for Retired People." That was 27 years ago, and the topics are still valid, as can be seen in the following description of the Managerial Grid Model.

Capitol Hill Workshop–April 1981

Another educational experience to keep us dumb engineers up to speed on how the other side of government worked. The theme of this workshop was: "Can government do the job it has to do?" *(plus ça change, plus c'est la même*–the more things change, the more they remain the same!–how true.) The intro stated:

"A new President, a Republican Senate, a more conservative House with 55 new members... all this provides a chance to make the new beginning that the Reagan Administration seeks. But the task will be difficult."

A massive housecleaning of government programs is needed if the budget is to be brought under control. The need for increased defense spending and a more assertive stance in foreign policy must be reconciled with the need to reduce over-all spending. Constituencies stand ready to fight every cut, and special interests must be brought into line with the public mood to alter the direction of government.

This seminar will be held in the middle of the budget cycle—after authorizing committees have made their suggestions for spending cuts, and just as the Congressional budget committees are to report out their proposals for spending and revenue ceilings. The battle will be on. The highest skills and cleverest tactics, backed by the President's exceptional ability to use the "bully pulpit" of the Oval Office, will be needed to assure a respectable measure of success.

Through the eyes of White House aides, key congressmen, political scholars and outside observers, this seminar will chart the policies and politics that should help to make government more responsive and productive, thus making the economy and the nation more stable and efficient.

The relationship between the Congress and the Executive Branch, the workings of Congress, economic policies including new modeling techniques and tax cuts, constituent pressures, the role of the press—all these issues and others will be discussed in detail." *(they were; it was enlightening–and things haven't changed one bit in 26 years.)*

My Charge to Ballistic Missile Defense Program Managers 11/3/81

For a banquet at the Ballistic Missile Defense Program Managers conference, I had the distinct pleasure of an invitation from the Undersecretary of the Army, the Honorable James Ambrose, to give the keynote speech at their banquet. I have taken the liberty of including it here–it was short and to the point:

1. The cooks, waiters and other staff personnel were able to set a

banquet before you within a cost target, in a timely fashion, hot and tasty. Had the meal merely been hot, or even hot and tasty, but over cost and after keeping you waiting an hour there would have been a lot of unhappy people in this room.

Your problem as managers is not dissimilar. If your system is produced to specs but is late and over budget, or worse yet, has to be accepted on waivers, you haven't done your job. If you have accepted a design from your designers, which works in DT-I but is gold plated and hogged out of solid bar stock you haven't looked at the producibility.

If you've wisely programmed money to look at the producibility of your system during AD and ED but diverted the PEP funds because you had costly technical problems to resolve, you've just pushed your problems on to the shoulders of the next PM to follow in your post.

If you've decided to produce a system that works but haven't done a cost driver analysis to see where you can bring the cost down, you haven't done your job.

If you use new technology components in your system which promises to give us a quantum improvement in effectiveness–Great! But have you considered if the manufacturing technology exists to make the item at a reasonable cost and in time to feed into the total assembly schedule?

2. We are constantly assaulted with cold, tasteless fare, served late and well above the price we bargained for-not in terms of what we eat, but in terms of the projects paraded by us. And there's always a slew of reasons of why the project is late and over cost and often doesn't work the way it should. My old, field first sergeant always told me: "Excuses and Alibis Represent Failure."

Well, the best way I know of to prevent those failures is to look at the total system and the total production process-top down; tear it apart, step by step, and then put it back together methodically, with someone who's had shop experience and dirt under his nails, to make sure it's put together right and the raw materials flow into your factory at one end, get processed and come out as an inspected, finished product at the other end.

Just remember that every time you sit down to a hot meal, well served-if your product isn't hot and on time and within cost–you haven't done your job.

The Brookings Institution (Williamsburg, VA)

"1982 Public Policy Conference for Government Executives"

From 24 January to 5 February in 1982, the Brookings Institution held this conference in nearby Williamsburg. Why bother to write about one more conference? Well, this conference was to be my education on preparing for retirement. It's understanding public policy that makes an executive effective, whether working for the government or helping potential contractors to understand how the U.S. government works.

What made this particular conference so worthwhile was the roster of attendees, and, being off-site in this idyllic locale, provided a rare opportunity to exchange views with other executives and to learn from them, as well as the speakers. For example, the participants included the Assistant Postmaster General, the Comptroller of the Currency from the First National Bank Region in Boston, the Director of the Ballistic Missile Defense Program, the Budget Officer of the U.S. Geological Survey, the Director of the V.A. Medical Center, and, most exciting, a "special agent" from the U.S. Secret Service.

After WORK–PLAY!
Part 5
Our Travels from 1969 to 1983

Dieses war der erste Streich (this was the first act of mischief, as Max & Moritz said)-not that we had not traveled before to Mount Snow for skiing, to Miami Beach to visit with Grandma Fusje. Sure, we had taken a cruise to Bermuda shortly after Edwin was born, leaving the children with their grandparents. And once the kids were old enough to ski, we made many trips over the Easter school break to ski in Vermont. We even changed the Hebrew calendar one year, celebrating *Pesach* (Passover) a week early, so that when our hosts at the Vermont chalet served us pancakes and waffles at breakfast to make sure we were warmed inside before hitting the slopes, we could all partake.

1973 the "first" trip back to Germany & the *Altstadt*

Our first "real" trip, overseas, wasn't until April 1973, when we visited London, Frankfurt, Seligenstadt, Salzburg, and Innsbruck. This was my first reunion with my cousin Hans and his wife, Marianne, to visit the *Elternhaus* on *Steinheimer Strasse* in Seligenstadt, and then visit *Östreich* on our Eurailpass. It was a nice opportunity to mix business with pleasure and taught my good wife that Yiddish is NOT the same as German.

Still and all, this reunion after 25 years was as though we had never left and we just continued our conversations where we had left off so many years earlier.

A GRATEFUL REFUGEE KID'S RECOLLECTIONS

We heard about the tribulations of half-Jewish, half-Catholic kids in a small village, after their mother had been torn from the family and sent to *Offenbach* (where she managed to write a cryptic postcard, the first letter of each line spelling HUNGER!)

Hans and Erich and their father, Adolf Thoma, never saw their mother, Irene, again. She perished in the concentration camp, just as my aunt Else, her husband, Emil, and my aunt Lenchen, her husband, Julius, did. From Frankfurt, after visiting my old *Synagoge* on *Freiherr von Stein Strasse* and the new Jewish Museum, where we had a Kosher lunch, it was time once more to board the train and head South to Baden-Baden, where Erich picked us up and took us to their home, high on a hill above Bad Herrenalb, a typical spa. His garden and the way he had his tools stored were so immaculate and orderly as to be beyond belief, but it was another nice, warm reunion with my younger cousin and his wife Erika.

In Salzburg we took time to visit the *Schloss* (castle) high on the hill, the *Marionetten Theater* and 17th Century *Baroque Dom* (cathedral) where Mozart played, and just strolled around doing all the touristy things after getting off the cable car from our trip up to the *Festung Hohe Salzburg*. One thing my good wife, Shirley learned on this trip was that you don't need so many clothes, especially when you travel by train, rushing from track to track to make connections, and I had to lift the suitcases into the overhead rack. Thus once we got to our hotel in Innsbruck we did some repacking and sent one big suitcase home. Since then we've learned to travel with a small roll-on suitcase each and a backpack, and have never wanted for anything.

In Innsbruck we did the usual tourist bit, walking the narrow streets, window shopping and picking up goodies for a light supper in our room, complete with a bottle of the local wine, and, to top it off-some real Austrian pastry and a jug of milk. Big mistake! Wine and milk don't mix well.

As a result I was quick to wake up in the middle of the night when I felt the bed shaking. At first I thought my good wife had put a coin into some bed vibrator, but when I saw the straps on the suitcase shake back and forth, and the closet door swinging, I felt that I had better call the desk and told them: "*Das Zimmer wackelt.*" (the room is shaking) To which the desk clerk replied-"*Ach, joa, wir haben ein kleines Erdbeben.*" (we're having a "small" earthquake.)

The Train Ride to Venice-Mestre

Sure enough, next morning our train to Venice was detoured from its normal route through *Udine*, the epicenter of the earthquake, through Czechoslovakia. When our train stopped just inside its border and troops came aboard each car to make sure we stayed in place, I asked one of the soldiers in German, if I could at least get off the train and buy a cup of coffee in the station. He thought that was OK and I walked over, asked at the coffee shop for a *Tshashka Kofye* and had the barmaid squeeze out a cup of coffee grounds for me. Not having any local currency, I gave her a dollar bill and figured that it was worth it.

Well, a little while later as we stood at the window waiting for the train to pull out, here came the barmaid with a big burly *Bulyak* on either side, scanning the windows till they spotted me. I was "invited" to step out, and ceremoniously she handed me some change in the local currency. Seems a dollar was more than they could accept. Ah, yes-the almighty dollar, as we were to find out again when we visited the Soviet Union a few years later in 1991.

When our train finally reached the station in la bella Venezia, we had that wonderful experience all first time visitors have, of stepping out of the train station, down the steps and to see before us the grand canal, with *vaporettos* (water buses) stopping every few minutes, to take you along the winding canal to the grand plaza and the Doge's palace-though one could walk there as well, through the narrow streets, lined with shops selling lace, glass, masks, pottery and food!

Our room was at the *Pensione Monna Lisa*, just down a side street from the station, very convenient to all the sights. It was a time for sitting at one of the cafes in the Piazza, listening to the dueling orchestras on each side of the piazza, visiting the *Camera di Cinque Ciento*, the 500 dignitaries who met in the Doge's palace when assembled, to feed the pigeons, take pictures and luxuriate in lunch on the deck of the famous Pitti Palace (later bought up by Starwood Hotels.)

We walked the narrow back streets, crossing canals, past the Bridge of Sighs, (so named for the prisoners walking from court to the prison, probably never to see the light of day again) visiting the *Ghetto Nuovo*, where a visit to the second floor *Synagoga* and the little shop where *Gianni Toso* created glass Rabbi figures before your eyes, was obligatory and wonderful. Naturally we opted to buy a figurine for each of the children and for us- treasured to this day, 30 years later. (Now he has a studio and shop on Baltimore's Reisterstown Road in Pikesville.)

1974 Holy Days in Israel

In October 1974 we took our first trip to Israel, with stops in London and Rome. For my first visit back to London after 35 years, and Shirley's first, it was a good thing to get a tour, the better to be able to appreciate all this beautiful, historic city has to offer. From London we flew on to Israel, seeing Tel Aviv from the water approach, as everyone aboard applauded. We quickly got to know Eli, our guide and our bus driver, who would be with us for the whole 10-day trip. Without their intelligent advice and guidance we would never have been able to cover the sights from Haifa and the Golan Heights in the North, through Jerusalem, Nazareth, and via *Massadah* and the Dead Sea to Eilat in the South.

One of the highlights of our trip was the fact that we arrived in Tel Aviv on *Erev Yom Kippur*, and had the chance to visit and pray in the Grand *Synagog* the next day, where, as honored guests, we were seated in front by the Aron Kodesh, right behind the Rabbi. At 3 pm he turned to us and said: "This time last year is when we were attacked by the Arabs." The cantor, recently arrived from Russia, made this visit even more than memorable.

I still remember visiting *Bet Alfa* and *Bet Shehan* to see the Synagogue and in typical tourist fashion, taking pictures of all the sights-until some children on a nearby hill started throwing stones at us. In my fascinated state I had forgotten that it was Shabbat. They reminded me. It was to be the first of many visits for us, but this was a good introduction.

Of course, it was also a chance to visit with my father's cousin, Rudi Baer and his wife Erika and to see some of their children. Rudi and Erika had left Germany early in the 1930s and given all their wealth to Kibbutz Hazoreah, which they helped to found, where their children grew up, and where Rudi ran the plastics plant until well into his 80's.

After the memorable visit to Israel, Rome might have been an afterthought, but dining at our hotel, the *Ambassadori Palazzo*, we learned that this was to be the last night for the opera, and they were doing Cavalleria and Pagliacci. Now this was an event not to be missed. Little did we know when we ordered our *due biglietti,* that this was not the Grand Opera, but the *Opera Picola*-the "little" opera.

After a wonderful dinner (all Italian dinners are wonderful-well, almost all) we took a taxi to the Opera. My Italian at that time was pretty meager and when the driver asked for *due ciento Lire*, I misunderstood and thought he asked for *due mille*, (2000) not 200 and, knowing that taxis cost more at night, I paid what amounted to about $15 at the time, and walked up the steps to the opera. Half way up I heard the cabby call me "Signore, Signore-you gavea me too mocha moanie." I was so touched by his honesty that I split the difference with him-another memorable moment. You see–not everyone is out to rip off tourists.

1975 My trip to Korea and Seoul food

A year later, in 1975, I had the chance to visit Seoul again, on behalf of AMSAA, to visit some of our troops to see how well our supplies were working for them. At that time, this meant a stopover in Tokyo, where Bob McGowan and I went to the typical local restaurant and we used the limit of my Japanese vocabulary. When the waitress asked: "Tempura?" (specialty of the house here), I merely answered: "Tempura, Hai!" (yes) so we did not go hungry.

Once in Seoul we were taken by jeep, off-road, out to a remote site. It was to be a critical lesson for us, learning how bumpy the ride could be for equipment we delivered to the field. In fact, our young escort, whose arm was in a sling due to a "combat football" injury, was in such pain as his arm swelled up inside his cast, that they had to cut off the cast when we got there. This is how we learned that any equipment shipped to the troops, even movie projectors for an evening's entertainment, had to have special packaging to absorb the vibration.

We were told of an electronic replacement unit that had been shipped in to Seoul and tested out perfectly, which, by the time it got to the field, was useless. An expensive way to learn the difference between low frequency, high amplitude vibrations to which equipment is exposed when transported by truck (where it usually has to be tied down so it doesn't bounce off the truck bed), and high frequency, low amplitude vibrations, such as one might encounter in aircraft transport.

1976 Berlin and a European CALS Conference

In May 1976 it was time once more to mix business with pleasure as I had a conference in Berlin. While participating in a CALS conference there (*Computer Aided Acquisition Logistics Support*), I had some time to visit Mückelsee, where my Onkel Werner sailed his boat back in the happy 1930's and to visit the opera to see one of my favorites-*Hoffmann's Erzählungen* or Tales of Hoffmann. Little did I realize how disappointed I would be in the new, ultra-avant garde staging, nudity and bizarre costumes, which had become common in Germany. Regrettably this craziness has not changed, and even singers with whom I speak now, in my reincarnation as an opera volunteer, complain about some of the inappropriate, stupid staging. Ah well, happily this insanity hasn't migrated here to the US or to Italy, the real home of opera. (Oops, it's now 2009 and we find it here & in Italy.)

From Berlin our tour took us by train to Nürnberg's gabled roofs and its wonderful museums inside the old walled city. 30 years later I still recall the *Fachwerk Museum* with its ancient tools and artifacts, displayed in the small rooms of this typical old world building with spiral staircases (a precursor of the larger, more spacious Mercer Museum we were to visit some 25 years later during a family reunion in Doylestown, in well-known Bucks County, Pennsylvania) and the *Kunsthistorisches Museum*, with its magnificent displays of paintings and sculpture, and-yes the delicious *Schwarzwälder Kirschtorte* (Black forest Cherry Cake) they served in their *Kafé*.

Next stop was *Rothenburg ob der Tauber* with its famous angled passages and beautiful town square and the "Nunnery turned into a museum," depicting life and living of the 1600's. Salzburg was our next stop, to revisit places we had enjoyed during our previous visit and some new ones. From there it was on to alt Wien, walking around the old town, visiting the *St. Stephans Kirche* with its steep roof, in the heart of town, going out to *Schönbrun* Castle and its fantastic vistas and lovely gardens, thrilling to the Spanish riding school performance and taking in all the works of art at the beautiful art museum. We even found time to attend an operetta and an opera on our last night, almost missing our train, because we had to fight for a taxi to the *Bahnhof*.

Now it was time for a re-visit of the Venice we had come to admire on our previous trip, to see new sights and second visit to places we had learned to love before. Rome afforded us a chance to tour the *Castell San Angelo*, to walk at leisure through the Appenine Hills and see the house Caesar had bought for 1200 Lira so many years ago, the *Colosseo*, the "typewriter monument" (as my uncle, Werner Glaser, called the *Monumento a Vittorio Emanuel* II) and to stroll through the vast expanse of the *Campidoglio* with the ruins of the Church of the Vestal Virgins.

Napoli (Naples) was next, affording a side trip to the excavated ruins of *Pompei*, ravaged so many years ago by the lava and ashes from the eruption of Mount Vesuvius and a boat ride to Capri to see its *Grotto Azura* or blue grotto, a bus ride along the narrow, winding, two-lane Amalfi drive overlooking the Tyrrhenian Sea several hundred meters below, to Sorrento for an evening of *al* fresco dining and songs typical of the area. We almost missed our train to Florence, our next stop, when we got to the *Stazione Ferroviaria* (train station) and found no listing for our train.

Fortunately we were early and after asking around, able to catch the *Metropolitana* to the *Stazione del Norte*, where our train would be departing. Interestingly, while waiting on the platform there, a young Czech train engineer was trying to get directions and could not make himself understood. So I asked him in German what he was looking for and then translated it to the Italians, so we got him off on a good start as well. We had planned to visit Florence again just because we liked it so much before and wanted to visit again, and then on to Interlaken, a small picturesque village in heart of the Swiss alps, that afforded us a chance to walk around the lake and take a train and cog railway up to the *Schilthorn* for a gondola ride to the top, and then a trip to the train station inside the famous *Matterhorn* to thrill at the view.

43-Susan's 1978 Introduction to the Great Cities of Europe

In April of '78 we had arranged the "grand tour" to London, Paris, Florence & Rome for Susan's 16th birthday, and taking along her friend Aymee. As jaded as young 16-year olds can be, they loved every minute of this trip, as we were able to show them places we had come to love on previous visits, and expose them to the wonderful local foods at each place, including buying crêpes filled with jam from street vendors, gelato in Firenze and all the ancient, splendid works of art that have withstood the test of time and still excite young and old today.

1979 Visit to Grand Canyon, Oak Creek Canyon & Tlaquipaqui

Next year in October 1979, we decided that it was time to see some of America's grand vistas and flew out to Denver to drive through Oak Creek Canyon on our way to Grand Canyon, stopping along the way to tour the arts and crafts village and galleries of Tlaquepaque not far from Sedona. We've been back to Tlaquepaque and stayed overnight in one of its many charming B&Bs just to have the opportunity to spend more time strolling through its quaint shops and streets.

An interesting aside about Grand Canyon. It was getting dark as we drove North from our ManTech conference in Phoenix. Even though our rented car was showing that we were nearly out of gas, we tried to make the fastest approach to Grand Canyon, by-passing Flagstaff (where we might have been more likely to find a gas station.)

So here we were on this lonely two-lane road, with virtually no other cars on the road when a sign said "Gas Station Ahead." But when we got there, it was boarded up. Same thing with the next one, until, lo and behold, there appeared a dimly lit Indian country store and gas pump, like a mirage. Happily we tanked up and bought more trinkets than we needed, out of gratitude for not running out of gas out there in the wilderness.

Oak Creek Canyon

Shortly thereafter we reached the Canyon entrance, and were directed to our log cabin, where we gratefully unpacked and after a brief walk, to try to see what we could see of the Canyon in the dark (nothing) we went to the main house for dinner. Good American fare had left us sleepy as we returned to our cabin to find that a cedar wood fire had been laid in our fireplace with the most delightful aroma wafting toward us as we approached. Needless to say, we slept well.

But imagine our excitement, tinged with a bit of horror, when we awoke the next morning and strolled over to the place where we had looked out the night before, to find that there was a sheer drop-off just a few feet away–and no railing! Nevertheless, the sights, the walking tour, through the woods flanking the canyon and through its trails, were a most memorable experience. And yet, in a way, Oak Creek Canyon, shown above, was somehow even more memorable. It is not as vast, but since you can drive right by its outcroppings and stop to walk up the multi-hued mountains (except during the heat of the day when snakes like to bask there) it is really quite beautiful and awesome.

A GRATEFUL REFUGEE KID'S RECOLLECTIONS

1980 *Tour d'Europe encore–France, Italia, Schweiz, de Nederland*

In May 1980 it was time for a more in-depth tour of France, with visits to Paris, Cannes, Juans-les-Pins and its beautiful, sandy beaches, to Nice, not to miss the home used by Pablo Picasso during his earlier, more stylized painting period, to Monte Carlo, and then by rail on to Venice, Vicenza, Verona, Frankfurt, Bad Homburg, Utrecht, and Amsterdam. Just reading this itinerary now makes me tired, but what a grand tour it was.

It's a good thing we took videos along our trips or they would all become a jumble of images. Our first stop in Paris, after dropping our luggage with Mme. Berthier, dans nos chambre, at the Hôtel d'Angleterre on the rue Jacob, was a walk down the *Champs Elyssées*.

From there we walked through the *Jardin de Tuileries*, passing the Louvre on the left, crossing the Seine to revisit my favorite museum, le Gâre d'Orsay. The old train station had been converted by women designers and architects, and whose *directice* was a woman, into what is now the magnificent *Musée d'Orsay*. In all my visits I've never tired of it. *The train station then and now (next pg)*

Of course, any walk down there, through the Plâce de la Concorde, always included a brief pause to visit the toilettes in the grande Hôtel Crillon. Dinner on our first night meant a visit to the colorful *Taverne du Sergeant Recruiter* on the *Isle St. Louis*–designed for tourists, but always a nostalgia trip.

And no visit to Paris is complete without a visit to Montmartre, eating outside on the *Plâce* where artists display their wares for the tourists, and an accordionist played *Musique typique*-typical Parisian French music.

A Eurailpass is a wonderful thing to let you get around Europe without the hassle of driving, and so it was easy to hop a train at *Gâre de Lyons* and head South to the *plâges* (beach) ahead of the vacationers who flood them in July and August. Riding the rails gives you a chance to see more of the countryside because you're not preoccupied keeping your eyes on the road, though it does preclude making a side trip to a little village that looks interesting. We'll do that some other time.

From the train station in Cannes it was an easy walk to a lovely pension just a short walk from the beach. And Cannes is, after all, well known as a splendid place to walk the boulevards, sightsee the expensive shops, admire the beauties of the passing parade and just luxuriate, when you've had enough of the beach for one day. Mme Yvette Geolle proved to be a gracious hostess, making sure that our accommodations were comfortable and steering us to places where the food was *typique,* tasty, *et pas tros cher.*(and not too expensive.)

The local train brought us a few days later to Juans-les-Pins, where the pebbly Cannes beaches were replaced by smooth sandy ones, and the sights, while not as luxurious, were just as beautiful and pulchritudinous. Our hotel, to which we were return time and again over the next few years, was just a few blocks down from the train station and a block from the beach. In fact, from our balcony, where we enjoyed our breakfast each day, we could look out over the beach–or admire the ladies who were not embarrassed to come out to their balconies for a morning stretch, in all their natural beauty–*en déshabillé.*

And there was the somewhat disconcerting event, as we sat at a table right in the sand, just meters from the beach, enjoying our *salad Nicoise,* when a well endowed lady, stopped to chat with a man at the table next to ours. It's just something we American men aren't used to–but can still enjoy with a sideward glance. Even my good wife giggled at my obvious discomfort.

Our next stop, heading East on a lovely May day in 1980, was Monaco with its splendid hanging gardens and refined, subdued atmosphere of elegance and luxury.

On to Italy

And so it was on to *la bella Italia,* which has, over the years and many more visits, become our favorite country to visit for its low-keyed beauty and the charm and hospitality of its amazingly pleasant populace. You know, you can find nice, friendly and helpful people wherever you travel, but somehow in Italy we've never run across anyone who didn't try to be helpful, no matter how fractured our Italian question came out.

In *Milano*, as in many old European cities, we found a lovely hotel just a few steps from the *stazione ferroviaria*. (note, that as in the French *chemin de fer*, or the German *Eisenbahn*–they all refer to the iron tracks or train.) We even checked out the hotel right within the structure of the train station, which was so well insulated and sound-proofed, that the raucous sounds of the trains and milling *Milanese* did not intrude into the quiet ambiance. Of course, it was not until years later that a thoughtful client put us up at the 5-star Le Meridien Excelsior Hotel Gallia Milan, just across the Piazza from the *stazione*– but that's another story.

We quickly learned to take the well-situated *Metropolitana* subway to the central piazza to see the *Duomo*, which, in those days, was still covered with dark stains from the coal and wood fires that warmed the populace in the winter. It was only many years later that its walls and hundreds of splendid statues adorning the walls, niches and roof would be scrubbed to reveal their original splendor.

From the *Piazza Reale* we walked through the block long covered market street to the *Piazza Scala*, to view the best-known opera house in the world, La Scala. We encountered an interesting, but disturbing experience on the Metro on the way into town. The cars were packed with flag-toting, rambunctious, noisy *passagieri*, who poured off the train as we got to the Duomo station and then formed up to parade around the square, waving their red banners and chanting noisily. It was a brief strike, so common in Italy, but disturbing to me in that it recalled the racous Nazi *Schutzstaffel* parades I witnessed in Frankfurt, as they roared their strident *Die Fahne hoch, die Reien fest geschloßen. SA marschiert mit ruhig festem Schritt."*

(Raise high the flag; close the ranks, SA marches with quite, solid tread.) I couldn't get away fast enough from these far less menacing marchers.

Our next stop was Verona, the town made famous by the story of Romeo and Juliet, but which incidentally, also had a fabulous museum, the *Museo di Castelvecchio*, This impressive castle built in the 14th century houses a fantastic art gallery right on the Adige river on the *lungadige*. Frommers' describes it so well, that I take the liberty of quoting him, so that maybe you too will want to visit this amazing museum:

"A 5-minute walk west of the Arena amphitheater on the Via Roma and nestled on the banks of the swift-flowing Adige River, the "Old Castle" is a crenellated fairy-tale pile of brick towers and turrets, protecting the bridge behind it. It was commissioned in 1354 by the Scaligeri warlord Cangrande II to serve the dual role of residential palace and military stronghold. It survived centuries of occupation by the Visconti family, the Serene Republic of Venice, and then Napoléon, only to be destroyed by the Germans during World War II bombing. Its painstaking restoration was initiated in 1958 by the acclaimed Venetian architect Carlos Scarpa, and it reopened in 1964. It is now a fascinating home to some 400 works of art. (Isn't it wonderful how so many bombed out buildings were restored.)

The ground-floor rooms, displaying statues and carvings of the Middle Ages, lead to alleyways, vaulted halls, multileveled floors, and stairs, all as architecturally arresting as the Venetian masterworks from the 14th to 18th centuries—notably those by Tintoretto, Tiepolo, Veronese, Bellini, and the Verona-born Pisanello—found throughout. Don't miss the large courtyard with the equestrian statue of the warlord Cangrande I (a copy can be seen at the family cemetery at the Arche Scaligeri) with a peculiar dragon's head affixed to his back (actually his armor's helmet, removed from his head and resting behind him).

A stroll across the pedestrian bridge behind the castle affords you a fine view of the castle, the Ponte Scaligeri (built in 1355 and also destroyed during World War II; it was reconstructed using the original materials), and the river's banks." You can see why we were so taken by this place, as we stopped on our way from a visit to the famous balcony, down tree-shaded lanes back to the railway.

Then it was on to Vicenza for a quick business discussion with MG McFadden. Those were both places we would visit again in years to come. But, for now, we wanted to get on to *la bella Venezia*. As I've mentioned before, I shall never forget coming out of the *stazione*, walking across the terrazzo and down the steps to see before us *Canale Grande* and one of the many stops for the *vaporetto*, the water bus, that takes you all over Venice. (now, reading the mysteries by Donna Leon, which all take place in and around Venice, it's nice to have some familiarity with its layout.)

It was just a short walk down the *Calle Priulli Cavaletti* to the left to reach the *Pensione Monna Lisa*, which had served as our base station before, for the next few days as we explored the shops and narrow alleys around the Piazza San Marco, with its *Camera di Cinque Ciento*, the grand room in the *Palazzo Ducale* where the 500 leading citizens used to meet to decide matters important to the welfare of the Venetians.

Like all the other tourists we fed the pigeons that abound in the Piazza, with its dueling orchestras playing waltzes in front of cafés on the north and south sides of the piazza, and, of course, stopped for some of their delicious *gelato* and coffee. Before heading to the Jewish quarter in the *Ghetto Vecchio*, we took pictures at the Bridge of Sighs, which convicted felons crossed onto their way to prison. It was time then to stop for lunch on the terrace of the famous Hotel Danielli, looking out over the *Canale di San Marco* to the magnificent *San Giorgio Maggiore* cathedral.

It was on the northern corner of the *Ghetto Nuovo*, across from the very old *Sinagoga,* that we returned to the little glass shop, where, as we peered in the window, Gianni Toso was busy creating little glass figurines of Rabbis and other bearded figures. As he waved us in, like old friends, we came in to stand, as on our previous visit, watching over his shoulder, awed by what we saw, and quick to purchase three different ones–one more for each of our children. We were to buy more from him in years to come, before he opened another shop in Baltimore.

Seeing a Bit of *Östreich, Deutschland and de Nederland*

Reluctant to leave this colorful, historic town that once ruled the world, we headed North to Innsbruck, Austria to spend some time at the Hotel Bellevue on the *Hungerberg*. True to its name, the hotel did provide *une belle vue* (a truly lovely view) over the town and surrounding countryside.

Of course, getting to the hotel meant going across town, past the famous *Goldene Dächel* to the cable car that would take us half way up the *Hungerberg*, from where it was just a short walk to the hotel.

Innsbruck is a charming, touristy town, where one can explore and when you are tired of the tourists, take a peaceful walk through the woods, like the natives. Our hotel, the Bellevue, was beautifully situated half-way up the *Hungerberg* with a marvelous view of the town below and the bucolic countryside. We hated to leave, but *Frankfurt-am-Main* awaited us, and we spent the next day there, with side trips to *Bad Homburg-vor-der-Höhe*, where my mother was born, and to visit the *Elternhaus* in that quaint old village of Seligenstadt, where my father was born.

From there we headed North to Bonn, when it was still Germany's capital, to visit our military establishment at Bad Godesberg. Then it was on to Utrecht in *de Nederland*, where you'd better not try to ask directions in German, as the Dutch will turn their back on you and walk away. Better to try English. Our stay at the *Cyrano de Bergerac on the Kromme Nieuwe Gracht* was charming and comfy.

Miami Beach, FL

The advantage of having Grandma Resnick living in Miami, was that it gave us a good excuse to drive down to Florida in December of 1980 and we got a chance to soak up some sunshine while it snowed back home.

It was to be our last opportunity to visit her there as she passed away a year later.

The 1980 & 1981 California MTAG* Conferences

Back to business! In January '80 we had the chance to spend some time in San Francisco, Palo Alto and Sausolito to participate in the 11th annual Tri-Service, *Manufacturing Technology Advisory Group* meeting in Monterey, California, arranged by the good folks at the Weapons Command at Rock Island, IL.

They always found some place where it was so pleasant that you hated to go into the conference hall for the meetings–unless, of course, you were giving the presentation. Needless to say, this venue allowed for lots of sightseeing in our spare time. This was actually MTAG '79, but scheduling in 1979 had gotten messed up.

Meeting on the West Coast, fortuitously gave me a chance to keep in touch with our son, Edwin, who had been posted to the Naval Language School in Monterey for a year's immersion course in Russian. To this day he still remembers our lunch at the Sardine Factory, sitting by the fountain, and our personal waiter who made sure that had chilled salad forks. Turns out that when he enlisted in the Navy, who were destined to instill in him some of the discipline he shunned throughout his college years of indulging, he ranked so high on their language aptitude test that they nabbed him right away. It was the beginning of "enlightenment" for him, which started him on a distinguished and exciting career.

Previous MTAG conferences had been thoughtfully scheduled in Philadelphia, Pennsylvania (1975,) Arlington, Texas (1976,) the 9th annual MTAG conference took place in Orlando, Florida in October 1977, with LTG Bob Baer as keynoter, and the 10th in San Diego, California in November 1978.)

The 12th Annual MTAG Conference was held in Bal Harbor, Florida, with a keynote address delivered by my boss, General Jack Guthrie, the Commanding General of the U.S. Army Materiel Command.

In May 81 we had a chance to visit Longwood Gardens with my cousin Hans and his wife Marianne. Aside from marveling at the splendor of the indoor and outdoor gardens, they enthused over the generous trappings at our roadside fast food places, like Burger King– which have since then become common throughout Asia and Europe. We even found one in Hong Kong looking out over Kowloon Harbor.

(Whenever we have visited Longwood Gardens or any arboretum of similar splendor, like the Ladeu Gardens or the splendid Butchart Gardens in Victoria, Canada, we always recall the comment of my late father-in-law, Morris Resnick. After visiting the Longwood Gardens glassed-in arboretum, he said that now he had some idea of what heaven might be like.)

1981 was another busy travel year, and August found us in Bruges, with its beautiful canals and quaint buildings, Ghent, and Lieges, Belgium. The purpose of this trip was a Production Readiness Review at FN (Fabrique National de l'Arme du Guerre) in Herstal, conducted as part of my responsibility as Director of Manufacturing Technology for Munitions and Weapons. Since our prior visit and the accidental discovery of their superior minimitrailleuse (later to become our M-16 rifle) FN had become a major supplier of weapons and technology for our army.

The 13th Annual 1981 MTAG Conference

December 1981 found us at another Manufacturing Technology Advisory Group Conference, again cleverly sited on the small touristy Coronado Island in San Diego Bay, that afforded us the opportunity to also visit Monterey, Carmel-by-the-Sea, San Diego, and La Jolla with its lovely park, overlooking the ocean. Nothing like returning to the "scene of previous crimes" (or pleasures), even if most of the days are spent inside conference halls to be educated on the latest advances in Manufacturing Methods and Technology. It was, nevertheless an excellent opportunity to learn, and for contractors to show their wares and latest technology.

Touring Europe Again in 1982–just for fun

In June of '82 we decided on another visit to our favorite spot on the French Riviera, Juan-les-Pins, for some much-needed R&R. Office politics, dealing with a staffer who preferred to sleep behind his partitions, and another at a subordinate Command, who thought that he could get away with submitting last year's progress report with this year's date on the cover, were taking their toll. Happily, both the Union and the Personnel Office came to my rescue and convinced the "goof-off" to resign, rather than being fired.

So, after a suitable sojourn at the plâge, admiring the human scene, we headed East to Milan, and then North to another favorite hotel in the hills of Cernobbio, with its awesome view over the Villa d'Este to Lago di Como. Here the sights were more of natural beauty, and the tranquility of a boat ride on the lake, prompted one 19th century writer to assert that his life would be complete to "see Como and die."

From here it was a short hop to Lugano, in the Italian-speaking canton of Ticino and then on to the real Schwitzer capitol of Zürich and the French-speaking Neuchatel. An interesting little adventure in Zürich–we had been walking along the Limmat Kai (the quay along the river, down the street from the Museum of Fine Arts and across from the Opera House) when we were accosted by some rowdy young folks, who were obviously strung out on drugs. When we mentioned this to our host at dinner, he noted that they had a problem with these young people who had bothered other tourists as well, but that when an unarmed policeman tried to discipline them, they simply picked him up and threw him in the river

In June of 1982 we took an extended trip to our favorite haunts along the French Riviera, driving South from Paris to Juan-les-Pins, a little beach town, where the hotel knew us and we were familiar with the area-the soft sandy beach, good restaurants, even beachside, where taking lunch was sometimes interrupted by topless ladies stopping to greet friends at nearby tables. Oy, such hardship.

Interesting side-note here. I was trying to be very *Frainch* and kept our passports, airline tickets Eurail passes, and travelers' checks in one of those pouches "typical Frenchmen" wear on their wrist.

On our last day in France I had made a promise to take my good wife to the 4-star restaurant of Roger Verget in Mougin-a-Haut (that's up the hill from Mougin au bord du Mer.) Took the train from Juan-les-Pins to Mougin and squeezed off the train at our destination. I had noticed the swarthy Arab who jostled me, but paid no attention.

All of a sudden, as we were leisurely strolling up the hill toward our lunch, I stopped and said "it's gone!" Well, you can guess the rest. The jostler had neatly snipped the strap of my pouch and there went all of our valuables-and our appetite. It's wonderful how understanding the *conducteurs* of the *chemins-de-fer* were when I explained the lack of our Eurailpass. They let us get back on the next train for Juan-les-Pins and so we packed up for the night ride to Milano-*senso bilietti* (without tickets.) Again the conductor accepted our story and we stopped in Nice to get temporary papers at our Consulate there.

Here, on one of the very few anti-American episodes we encountered, two taxi drivers refused to take us to *Ambassade Americaine* (American Embassy.) The third one took us only because I just gave him the address, and (after one of the other taxi drivers yelled to him that we were ugly Americans) let us off a block from the consulate, which by then, was closed. Fortunately the consul was just getting ready to go home and she let us in, commiserated over our story and gave us temporary papers to get to Italy.

Once more, after crossing the border, when we explained to the customs agent and the *condottore* "*mio passoporte e rubato,*" (my passport was stolen) their only question was "*dove?*" (where?) Learning that they had been stolen in France and not in Italy, they smiled with relief. Italy's honor had been preserved. (Little was I to know that Milano rivaled Napoli for its thievery-but more about that later.) And so we made it to Milano, got new passport photos and were quickly issued new American passports at our consulate there. (it was lot easier then!)

A GRATEFUL REFUGEE KID'S RECOLLECTIONS

We quickly learned how to get around on the wonderful *Metropolitana*, which took us to the magnificent Duomo, (whose smoke blackened exterior has since been scrubbed to its original splendor) and the covered shopping arcade that leads to La Scala at its other end. As happens so often in Italy, there was again a 3-hour strike, which we witnessed first hand, as sometimes rowdy workers with red armbands or bandanas, shouted noisily, and after debarking at the Duomo, proceeded on a raucous parade. Again, I couldn't wait to get out of their way as it was too eerily reminiscent of the parades I had witnessed as a child in Frankfurt, as Stormtroopers of the SA, the *Sturmabteilung*, or worse yet, the SS, Hitler's protective *Schutzstaffeln*, stormed through the streets, bashing Jews out of their way and creating indiscriminate terror (as also previously noted.)

Once we had reached the safety of the arcade we relaxed, found time to shop and even get opera tickets for a grand visit to *LaScala*, probably one of the most notable opera venues in the world.

Happily, *LaScala* did not suffer the fire that burned *La Fenice*, the classic opera house in Venice, to the ground. (you can read all about that fire, and the problems of dousing it, when fireboats couldn't get close, because the canal in front of the opera house had been drained for repairs, in Donna Leon's *"Death at La Fenice."*)

From Milan we headed North to spend a few days at a charming inn in Chernobio, high in the hills overlooking Lago di Como. That gave us a chance to stroll the grounds of the famous and luxurious Villa d'Este right on the lake, take a boat ride, stopping for lunch and a post-prandial stroll *lungarno di lago*. From there we took the train to Lugano, and on to Zürich, Neuchatel, Pilatus, Luzern, and on to Bern. These trips to Europe ultimately inspired us to come back again and again to visit places we had enjoyed and where we felt at home.

Summer of '83 in New England

You'd think that our life consisted of taking trips and having fun, but, let's face it, there is much wisdom in the old saw that "all work and no joy, makes Jack a dull boy." So, in July of '83 we headed North to

New England once more. Our first stop was in the village of Marblehead, where we had booked a room in a modest B&B right on the main street. The place had been recommended by our friends, General Al and Ruth Ungerleider, as being ably run by their son's mother-in-law.

Well, their tip was right on the mark–the place was ideally located to be in easy walking distance of all the tourist spots, and especially as the start of many lovely walks along paths bordered by splendid gardens on one side and the ocean on the other. The local restaurants were all great–good local food, pleasant service–what more could one ask for?

We hated to leave there, but we had made reservations further North at another recommended site–"La Domaine" in Bar Harbor, Maine. La Domaine, as the name might imply, was run by two French ladies, who spent the off-season in France, returning to their roots, and then came back to immerse their guests in a true French-style country Inn. Between the décor and the food, one felt that one was in a bit of France. And it afforded easy trips to Bar Harbor, and to ascend Mt Battie, both for the views, and to pick luscious, ripe blueberries or Mertille.

Summer Business in Augsburg, Milan, Zürich, Pilatus, Luzern, and Bern

Shortly after our return from New England, business called me to Augsburg to visit the Messerschmidt-Bölkow-Blohm factory to learn more about the way they performed the milling of Titanium aircraft components. It was a tricky operation in that they used a matched set of milling cutters to remove some 80 to 90% of the solid to come up with the desired shape–as light as possible. One of the things I learned from MBB was that when one of the cutters broke, they removed the whole set and replaced it with another matched set.

In the meantime they take the broken cutter and grind it, then after carefully measuring it, assemble it with three cutters of the same dimensions. That was a very worthwhile lesson.

From there it was on to Milan to visit three robot manufacturers to see which one we could use at our Tank Command in Warren, Michigan. One of the three had a sales and tech support works already in Detroit, and one awed us by showing that they had built the world's largest robotic device, as tall as a two story building.

Since we had further business in Switzerland, we took the opportunity to make a stop to visit Mount Pilatus and visit Luzern, before heading for the financial capital of Bern, where I briefed one of our contractors and made sure to visit the Baeren for which Bern was famous.

1983 "Retirement/*le Retrêt?*"–beginning a new career

When my 55th birthday rolled around and I had accumulated 31 years of Government service and the daily irritations no longer made it all worthwhile, I decided to retire. Except that the kindly rep in the Personnel Office said: "No."

"You have to wait till next week to get your best pension coverage." Well, who was I to refuse such good advice. I could stick it out for a few more days.

(like the little boy who kept scratching himself, until the teacher finally sent him to the principal's office. In a little while he came back with his fly undone and exposing himself. When the horrified teacher asked what he thought he was doing, he said that the principal had told him that he should stick it out till the end of the class, and then he could go home.)

Hans & cousins Peter & Mucki Euphrat in front of Oberweg 50 in 1936

Hans & cousins Peter & Mucki in 1938

Baer family at Elternhaus (the family home)
The Baer Family - the 10 surviving siblings

Hans & Schul tüte, the traditional school kit

Buchenwald letter to Hilde Baer

Buchenwald" photo of Berthold Baer)
On the left Berthold Baer, 1936; on the right December 1938, after his release from Buchenwald

Letter to Buchenwald Commandant asking for Berthold Baer's release

John as waterfront counselor at camp

John as riding counselor at camp

John fencing at college

John's wedding photo

Figure 6. Photographs of Foreign Plastic Blank Ammunition

1. Norwegian Caliber .306 Bakelittfabriken Løspatroner
2. French Caliber .306 Gevellot Cartouche à Blanc
3. German 7.62mm D.A.G. Plaz-patrone
4. French 9mm Gevellot Cartouche à Blanc

blank cartridges and blank cartridge adapter

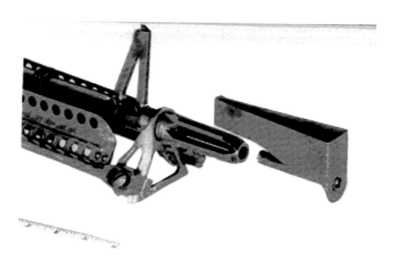

John & Col. Isenson at APG starting
paper recycling project

LWL CO Col. McEvoy

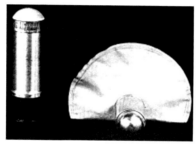

floating grenade

LWL secretary

Floating grenade in use in Vietnam by our troops to signal

JLB and truck armored with wood and glass-plastic windows

gun barrel engraving at Watervliet Arsenal in 1890

gun barrel engraving in 1980

Taiwanese worker sweeping foundry floor

John & Shirley at Welcome sign at Osaka foundry, Japan

Osaka foundry line overview

John & Shirley with foundry CEO Dr. Inaba

ISRAEL MILITARY INDUSTRIES
HAIFA DIVISION

John Larry Baer, President
INTERNATIONAL MANAGEMENT & ENGINEERING CONSULTANTS
4206 Elizabeth Lane
Annandale, VA 22003
U S A

October 19, 1986
Ref. RD/jc/BL

Dear John,

It with great pleasure that I take this opportunity to write a few words of thanks and appreciation to you, combining them with best wishes for a Happy, Healthy and Prosperous New Year.

Now that several weeks have passed since the proposal was completed, and we are better able to objectively review that period of hectic activity, we are able to appreciate all that you and Norm did to get a professional piece of work put together from disparate sources and organized into a cohesive, comprehensive whole.

We look forward to working together with you in the future and until such time remain

Yours sincerely,

R. Dotan

Razi Dotan, Manager
Bridging Systems Plant, IMI

HAIFA 31003, ISRAEL • P.O.B. 437 • PHONE (04) 549211 • TELEX 46415 • FAX (04) 537585

letter to IMI Export Office chief

International Management & Engineering Consultants
Advisor to Tech Evenement/ILCE, Paris
Member of Kramer Associates
4206 Elizabeth Lane, Annandale, VA 22003-3652 USA

John Larry Baer	David S. Bettny	Naomi McAfee	Marc R. D'Alleyrand	David Froom	Marc Pallot
President	Senior Partner	Senior Partner	Senior Associate	Senior Associate	Associate
Phone 703-323-6952	Phone 703-759-3882	Phone 430-744-6572	Phone 201-762-0614	Phone 44-753-885-677	33-1-45-864821
FAX 703-323-9845	FAX 703-759-3595	FAX 410-744-6572	FAX 201-762-0813	FAX 44-753-885-677	33-1-39-498342
jbaer@cap.gmu.edu	bettny@cme.nist.gov			david@fac.demon.co.uk	Paris Office

FACSIMILE TRANSMISSION SHEET

How does this look?

FROM: JOHN LARRY BAER, P.E.

NUMBER OF PAGES (INCLUDING THIS COVER SHEET)   __1__

DATE: 1 septembre 1994   FILE: CALS94\Pallot.901

Marc Pallot, ESoCE
18, rue Sthrau   TÉLÉFON: 011-33-1-45 864 821
75013 PARIS, FRANCE   TÉLÉFAX: 011-33-1 39 498 342

Dear Marc:

Thank you for your FAX. Let me address each point. 1) Even though you're approaching 40, this is not a cause for "un crise." There are always challenges ahead and you should never abandon your dreams. Look at me. I turned 65 last year but I'm still active and have no intention of slowing down.

2) Don't forsake your dreams. a) I read that the situation is picking up in France. b) I suggest you contact Walter Müller at Daimler Benz (Tel. 49-711 179 2869; FAX 4111) because he & his chef Dieter Hege, are looking to implement CALS/CE and you might be able to help them.

3) Don't ever hesitate to contact me when you need some paternal advice. That's what I'm here for. In fact, if you'd like we can put you on our letterhead as an associate, if you think it might help you. Then you could claim all the support of IMEC as you try to implement CE and your MEI notions. Let me know. (Shirley and I have many "children" of your age (not just our own three), who frequently call for advice.)

4) I've asked NSIA to check once more about your paper. I know Ron Taylor is looking for "foreign" papers & I will call him again too. Nevertheless, we are prepared to schedule an MEI workshop for you Tuesday evening and put it on the agenda (you're right about Friday as being "after the fact.") I will also try to contact John Halpin.

5) What the hell happened to Hans Günter Thonemann? Is his back still giving him grief or is he just too busy to answer phone calls and letters? Also, when you next see him could you get a proper e-mail address for him. The one he gave us doesn't work.

Please let me know tout-de-suite whether you want your CE workshop for 2 hours Tuesday night, Dec. 6th as I have to put in a room request for you. We will assume 50 people and conference style set-up. Naomi McAfee will **not** be at CALS Expo; she sends you her regards.

Sincerely,

John

European Society of Concurrent Engineering

TO:	John BAER	
company:	IMEC	fax: (19.1) 703 323-9045
FROM:	Marc PALLOT	
telephone:	(33) 1 45.86.48.21	fax: (33) 1 39.49.83.42

Date : 2 Septembre 1994   Page(s) : 1

Dear John,

Thanks for your "paternal" advice John. I have appreciated your messages and it is going directly to my heart.
You are, also very kind to offer me your support and I know that I am already using it when I ask you many things. Sometime, I am really thinking if I do not exaggerate asking you so much.

I agree with your proposal for the "MEI workshop meeting" on tuesday evening for two hours. I guess it should be between 5 and 7pm. It is OK for a room request. I hope that writing this meeting on the agenda will help to have about 50 people interested about integration prospective.
Please, let me know if it could be possible to have article about the Paris MEI workshop results in some newspapers like CALS Journal before the meeting to interest and atract people to come to the MEI meeting on December 6th evening.

Coordinates of Dr. Horst Soboll from Daimler-Benz are :
Dr. Horst Soboll - Daimler-Benz AG- AIT Project Office - 70546 Stuttgart
Tel : 49 711 17 92939 - Fax 94857

If I meet Hans during CALS Europe I will ask him a correct e-mail address but I am affraid that exchange with Hans are episodics. I mean it seems difficult to contact him most of the time and sometime he appears from nowhere and then he desappears again ...

I am very honoured to see my name on IMEC letterhead, it looks like very nice. And if you need anything from France or Europe let me know, I will be pleased to get it for you.

AIT project is a large Esprit III - CEC project involving Automotive and Aerospace industries in Europe and means "Advance Information Technology".

Yours Friendly,

Marc Pallot

Kiev soldiers' memorial to WW II

Государственный комитет СССР по стандартам

ВСЕСОЮЗНЫЙ ИНСТИТУТ ПОВЫШЕНИЯ КВАЛИФИКАЦИИ
РУКОВОДЯЩИХ И ИНЖЕНЕРНО-ТЕХНИЧЕСКИХ РАБОТНИКОВ В ОБЛАСТИ
СТАНДАРТИЗАЦИИ, КАЧЕСТВА ПРОДУКЦИИ И МЕТРОЛОГИИ

КИЕВСКИЙ ФИЛИАЛ

252115, г.Киев
ул. Святошинская, 2

Расчетный счет — 263611 в Ленинградском отделении Госбанка г.Киева

6.12.1990
№
На № ____ от ____

to: ROBERT L. CHARLTON
president
National Capital Chapter
Institute of Industrial Engineers

Dear Sir,

I'm sorry for troubling you, but we need your advise badly. The talks we had in Kiev at our institute, made us change our minds as to the problem of training our specialists in this country.

We would like to train our specialists who would meet the world requirements in the sphere of quality control, metrology, standardization using the technical basis of our institute, and after graduating these courses to give such specialists not the USSR Statestandard certificates but the certificates of international type.

There are no such centres in the USSR, but there exists a great need for such specialists. Many enterprises are ready to pay hard currency for such training of specialists.

To our minds to organize such centre we'll need:

1. participation of teachers-specialists from abroad;

2. special training for a number of our young teachers abroad and getting qualification and knowledge by them necessary to carry out the teaching process in the USSR. With this aim in view, we would like to get financial support from some American or Canadian funds. Unfortunately we don't know whom to address;

3. to organize on the basis of our institute the branch of international organization for training specialists of the enterprises from the Ukraine, Byelorussia, Moldova, South-East of Russia, the Caucususus and Middle Asia (these are the regions we are working with).

I will be very grateful if you help us to solve these problems.

Thank you in advance.

Yours sincerely
Dep.Dir. of Sciences
Dr. V.N. Novikov

ППП УкрНИИНТИ. Зак. 2468а-2х5000.

Russian letter asking for assistance

Kiev civilian memorial to WW II

The Great Gate of Kiev

Vu graphs used in Brazil briefing

photo of JLB speaking in Athens

disgruntled JLB, waiting for Greek Minister to finish very long diatribe

October 28-31, 1996
Long Beach Convention Center
Long Beach, California

Commercial • International • Government

Pour Delphine CLAUDON, le Secretaire de
M. Henri Martre, GIFAS & GIC-France
Honorary Chairman, Aerospatiale
115, rue de Bellevue
92100 BOULOGNE, France

FILE: CALS96\Martre.726

TÉLÉFON: 011-33-1-4699 0364
TÉLÉFAX: 011-33-1-4699 0361

Madame:

    For the Tuesday, 30 July meeting of the CALS Industry Steering Group which begins at 1 PM, we are setting up a Conference Call at 2 PM EDT (8 PM Paris Time.)

    If Monsieur Martre would like to call in and participate in this session by phone, we would be happy to provide him the appropriate access procedure and number. We have done this in the past with David Froome, UK CIC, so this is no problem.

    There is, at the present time no plan for a conference call for the 8:30 AM IBOD meeting, to be chaired by Aris Melissaratos, but that can be arranged if you let us know your wishes.

Je vous remerciez pour votre assistance.

Amicalement,

JOHN LARRY BAER, P.E.
CALS Expo '96 Plenary Chair

National Security Industrial Association (NSIA)
1025 Connecticut Avenue, N.W. • Suite 300 • Washington, D.C. 20036 • Tel. (202) 775-1440 • Fax (202) 775-1309
email: calsnsia@delphi.com

International Management & Engineering Consultants
Advisor to Tech Evenement, Paris
4206 Elizabeth Lane, Annandale, VA 22003-3652 USA

John Larry Baer	David S. Bettry	Naomi Schäfer	Marc R. D'Alleyrand	David Procme
President	Senior Partner	Senior Associate	Senior Associate	Senior Associate
Phone 703-323-6952	Phone 703-759-2882	Phone 410-744-6572	Phone 201-762-0614	Phone 44-1628-523-120
FAX 703-323-9045	FAX 703-759-3595	FAX 410-744-9686	FAX 201-762-0813	FAX 44-1628-526-718
jbaer@capaccess.org	bettyd@erols.com	ghajncafee@aol.com	infoxfer@aol.com	david@dffac.demon.co.uk

FACSIMILE TRANSMISSION SHEET

FICHIER: IMEC97\Transpor.113

Messieurs ou Mesdames:

En L'Usine Nouvelle du 19 Décembre 1996 j'ai lu cette vous proposer plus des réunions en le sujet du Transport et logistique, particulièrement en Düsseldorf, Madrid, Paris, Bruxelles, etc.

Si il est possible, voulez vous m'envoi plus d'information pour ces réunion. Merci très beaucoup pour vôtre assistance.

Veuillez agrée, Monsieur ou Madame, l'expression de nos salutations distingué

JOHN LARRY BAER, P.E., Chair
CALS Expo International '96 Plenary and
CALS Expo '97 International Sub-Committee

picture of storm destroyed
Tegucigalpa bridge in Honduras

Truck that was swept through the roof and boiler inundated in mud in Tegucigalpa factory

John with former president of Costa Rica

Part 6
International Management & Engineering Consultants (1983-2003)

I retired on Friday, the 4th of September in 1983, and my phone started ringing Monday morning. I got a call to come to IDA, the Institute for Defense Analysis, to work on an important task–an "Export Control Project." I had already read and signed their Consulting Agreement the previous week, and, along with some very distinguished people, served IDA in their Scientific Technology Division for several years–when I wasn't called to work for others.

Ending 1983 at the Institute for Defense Analysis

So my next assignment (which actually started in August of 1983) was working for the Institute for Defense Analysis on their "Export Control Project." As one of some 35 consultants, I had to apply my experience and expertise in arms and armaments, garnered over 30 years with the Army Materiel Command, to try to define what technology might or might not be exported to friendly and "not-so-friendly" countries.

A part of this work was to conduct industrial sector analyses in the area of computer controlled machine tools and robotics in Germany, Italy, Switzerland and Spain. Unclassified aspects of these analyses found their way into print in such articles as "Improving Productivity through Manufacturing Technology," "Engineering for Producibility,"

& "Foreign Dependency in Military Purchasing" in the Army R&D Magazine. Other works that came out of this, were a "Critical Technology Classification Guide," a 1985 report on "The Other Side of Outsourcing," and a 1986 report on "Foreign Dependency in Military Purchasing."

Next I got a call from Dr. Marvin Cetron, the head of Forecasting International, whom I had met at some of the Manufacturing Technology conferences we ran. He invited me to come down to his office, where he told me about a contract he just received from General Motors Central Foundry Division to investigate best foundry practices around the world-now and for the future.

What this involved was a 10 week tour of all the major foundries around the world to find out for the GM CFD how they could make better castings, and to do so without the objectionable flash they were getting, that demanded so much work to remove. Talk about an assignment made in heaven!

After visiting Central Foundry and learning the extent of their problems, in April 1984, I planned which foundries to visit and wrote each one, asking for an appointment, on a schedule roughly planned to take me to Europe in June, then on to Israel and Asia, and finishing in Tokyo in August. That was quite a trip.

I had already told my good wife that I was only going to accept jobs that would be fun. Her typical, woman's reaction was: "You're crazy-you can't do that." My reply was: "Watch Me." It was a vow I was to carry out several times during the ensuing twenty years, when, reluctantly and diplomatically, I had to tell a client that I could not work with them, usually because their business practices or personalities were just too unsavory and unacceptable for me.

Our first non-business vacation

But first-we decided to treat ourselves to a motor trip through France and Spain. So, in September, the shoulder season for travel in Europe. we took off for Paris, stopping in *Reims* (they pronounce it

Rähns) to view le Cathedral, and then headed South for a stay at the Hotel *Ombremont* on the lovely *Lac du Bourget*.

We had actually had reservations at another place, which shall remain unnamed. But when the concierge escorted us to our room, she looked in and immediately closed the door, saying: *"Je regret, Monsieur/ Une person a mort dans cette chambre."* (I'm sorry, Sir. Someone died in that room.) So she took us to another room, which looked morbid to us and had a mattress that was soft as mud, and we trooped back down the steps, and told the young lady: *"Je suis desolez*; (I'm sorry) we spoke to friends of ours who invited us to stay with them." And we took off for the *Ombremont*, situated in a beautiful garden and facing a lovely lake.

Our stay there was absolutely idyllic. Lovely walks, good food, the company of a friendly dog and house cat, and a *platte des fromages* (a huge platter of assorted cheeses) after dinner that was *vraiment Incroyable* (truly incredible).

We have never before or since seen, smelled, tasted and delighted in such a formidable array of cheeses (*Fromages de Savoie*)-each one better than the other. We hated to leave the place, but promised ourselves to come back another year.

So we headed further South through *Talloires*, crossing the Swiss border to *Lausanne* and *Chillon* and then up the mountain via *Chamonix* to Mont Blanc. The sights of the mountains, the lakes, the chalets and just the rolling Alpine countryside, were, of course, spectacular. Then it was back via *Oranges* and *Nimes* down to Spain, stopping in *Tossa de Mar*, South of *Rosas* on the *Costa Brava*.

That involved getting off the auto route/*Autostrada*, driving over a mountain, down toward the shore. As we entered *Tossa de Mar* we found ourselves on a narrow lane, just about wide enough for the car to pass, as pedestrians jumped up on the narrow sidewalks.

And then, all of a sudden, the street opened up as the beach lay before us. Making a turn to the right, toward the castle high above us, we found the parking area for the *plaja* (beach) and the little hotel that faced the beach and bay. How spectacular can you get?

This was a place that didn't invite you to relax; it said "Sit! Stay!" and so we walked the beach, strolled along the little streets and augmented our window shopping with the real thing, ate, slept, snoozed, and just had a hell of a lot of fun and relaxation. Then, since we were expected at my mother's cousin's place in Barcelona, we left, most reluctantly, and headed there. Lütt had left Germany in 1934, with her sister Hertel, after Hertel had the audacity to slap a Hitler youth who tried to kiss her. Emigrating to Spain meant conversion to Catholicism, to which Lütt adheres to this day, albeit, somewhat half-heartedly.

Driving into Barcelona is always a hoot, no matter from which side you enter. It's broad boulevards, plazas with huge fountains and well marked streets allow easy access, even for *stranieri* (strangers.) In fact, after leaving the apartment, which Lütt had occupied for some 50 years on the *Calle Sagues*, and regretfully giving up our hard won parking space, we could not get to our hotel. So I pulled up by a motorcycle cop and asked him: *"por favor, yo quiero il via por esto hotel. Donde es?"* (please, how do we get to this hotel?) And he simply said: "follow me" and led us via several one-way *Calles* up to its front door.

(*Which reminds me of a cute story about some Americans who were on a visit to Munich and, having sampled a few too many glasses of the Bock Beer, decided to take a taxi back to their hotel. The problem was that they couldn't remember the name of their hotel, nor even the street–till one of the gang told the taxi driver–"I remember–it was called Einbahnstraße" or One Way Street!*)

Next day Lütt led us up Montserrat, where we watched a spectacular parade of religious mannequins on stilts. Then it was time for lunch at a famous waterfront restaurant and a visit to the Queen's private residence, donated to a Nunnery, which now displayed a marvelous set of shadow boxes depicting various religious scenes, both from the old and new testament.

When Lütt asked when we wanted to go to dinner, we naturally said about 7 o'clock. After she got through laughing, she made reservations for 9 PM, when her favorite gourmet restaurant opened. Well, when we got there, we were the only patrons and had a hard time prying the chef and serving staff away from their tables.

Nevertheless, we had a scrumptious dinner, and by the time we finished, around 11 PM, the restaurant started to fill up, including families with children. We never did get used to the late hour. It was a grand finale to our trip to Spain.

A 10-week Dream Assignment

A 10 week Dream Assignment for General Motors Central Foundry Division–seeking out
"Best Foundry Practices" in England, Sweden, Germany, France, Spain, Portugal, Israel, Hong Kong, Taiwan, Korea and Japan for Dr. Marvin Cetron, Forecasting International, resulting in *"Foundry 2000-A Technology Forecast"*

Now this is where it gets interesting-Marv Cetron set up a meeting for us with the Director of GM's Central Foundry Division in Saginaw, Michigan, who took us on a tour of the facility, and whose basic instruction to me was: "Go find out what other foundries around the world are doing better than GM, what *avant garde* technologies they are using, and maybe how we can stop getting flash around our castings, which is a nuisance, and costly to remove."

Had I had this challenge today, it would have been fairly simple to do a Google search for the world's leading foundries. But, twenty years ago in 1984 we had to rely on the Thomas Register to get our information and to create the list of foundries we would visit in the months to come. First, three weeks later, we paid a visit to GM's Aluminum foundry in Bedford, Indiana, and then in June scoured the Robotics Expo for examples of how robotics could be used to ease the labor intensive foundry work. (Remember this, when we talk about Swedish foundries.)

While in Detroit we took the opportunity to also visit the US Army Tank Command in Warren, where my old friend, Don Cargo, was still running the Manufacturing Technology program, the GM Tech Center in Detroit and Marco Possati's robot firm, called MARPOSS. I had learned during my previous visit to Milan that the Italians built some of the best robotic devices in the world, and some of the most humungous forging equipment. So, after carefully culling the list of foundries all over the globe, we started in early May writing letters to those with the best reputation and very quickly received invitations to come to visit. It was for them a chance to let the world hear about them.

On Sunday, June 10th, 1984 we took off for Europe and arrived in Gatwick early the next day. Monday, a pleasant train ride brought us to Deritend Investment Castings in the Worcestershire town of Droitwich, about 150 miles North of London. We found a heavily robotized facility turning out top quality aluminum castings, as we did the following day at Cosworth R&D Ltd.

The surprising thing for us was that our hosts, besides being eager to show what they did, made arrangements for us every step of the way, from train schedules and taxi schedules to hotel reservations. (I know, I keep saying "us" because my good wife accompanied me on the trip, serving as amenosienses and corporate memory.)

I had written to the Director of Cosworth that we were "doing a world-wide study of the casting industry… to highlight its strengths… and to identify trends in emerging technologies."

Director, Cosworth Research & Development Ltd
Ten Acres, Berry Hill Industrial Estate
Droitwich, Worcestershire, England   26 May 1984
Dear Sir:

Based upon the suggestion and kind assistance of Mr. Thomas, Director of the Automobile Iron founders I look forward to the pleasure of visiting your facility on June 12th. So that you will better understand the reason for my coming, permit me to explain.

Under the aegis of Dr. Marvin Cetron, president of Forecasting International, Arlington, VA, I am doing a worldwide study of the casting industry, with particular emphasis on the future of iron and aluminum foundries concerned with the manufacture of engines and other vehicle components.

My aim will be to try to highlight the strengths of what is generally believed to be an industry in trouble and to identify trends in emerging technologies of this and other industries, which might help the casting industry out of its current doldrums. I will be particularly concerned with trying to identify technologies, skills and advanced equipment used in the new high tech industries, which might be usefully applied to enhance the casting technology over the next 20 to 25 years. As a professional engineer with over 30 years of experience in government and industry I am, of course, honor bound to protect any proprietary in format ion which I may receive and will restrict anything I write or publish in a report to those general principles which you agree to as permissible.

As I told Mr. Thomas, I expect to arrive at Gatwick on 11 June at 6:50 AM and spend the 11th at Deritend, staying over night in Worcester or Droitwich. I can be at your plant any time on the morning of the 12th and will probably have to catch the 2:48 train out of Birmingham in order to get to London in time for my 6:38 Heathrow departure for Stockholm. Looking forward to my visit to your plant, I am,

Very truly yours, John Larry Baer, P.E.

The major finding at Cosworth was their work in aluminum engine castings, using lighter, thin-walled and more complex components integrated into the basic casting, rather than being bolted on. As a result Cosworth found that the mechanical properties of their aluminum header and engine castings were 40% higher than for conventional castings.

Our "findings" at Deritend Aluminium Die Castings, Ltd., Blackpole, Worcester

Deritend had closed their sand casting factory in 1983, since they no longer found it commercially viable due to shrinking market for Nickel base alloy castings. They consider themselves a small flexible job shop who specialize instead in lost wax investment castings for 500 to 600 highly accurate missile & aerospace parts, torpedo motors, navigation and military communications equipment castings. They also make frames for the Marconi heads-up aircraft display. Deritend uses CAD (computer aided design) for the component to be processed but not for the gating system since they were unaware of the existence of such a program.

It took them four hours to make a wax pattern which is framed carefully in plastic and glued to minimize the risk of bumping and deforming. The pattern is then placed on a square edged surface to reduce the hazard of wax slumping under its own weight. It is then given the first of two ceramic coatings using a water-base slurry. The first coat is considered most critical in determining the finish of the part; the second for strength. This step is followed by seven coats using an alcohol-base slurry, building up the coating to about 1/4" and drying between each coating. Finally the mold is given a sand spray coating.

These 3-minute cycle time dip and spray operations, which take a total of five hours to apply, are an ideal application for the highly articulated robot they use. One man using this robot now does work formerly done by nine men. In fact, Deritend plans to use 3 robots in an integrated system to transfer pieces from the assembly line to a conveyor and then load them onto tracks. The pattern is air-dried on the shelf for 3 days to obtain green strength and then baked for fire strength. 70% of the wax core is then melted out at low temperature and the remaining 30% at higher temperature as the mold is baked.

The process, which is highly labor intensive, results in a very accurate but costly casting. After pouring and cooling the aluminum part the mold is removed with high-pressure water jets rather than vibration. A second water blast cleans any remaining residue from the inside of the casting. Feeders are removed by hand grinding and, of course, melted down for re-use. These castings in the first step of a 2-step heat treatment, are held for 15-hours high temperature solution heat treatment near their softening point and then quenched in liquid, which causes them to distort considerably, but is necessary for proper strength characteristics.

All castings are checked in a gage and straightened manually, using hammers. On larger pieces one man will stand on it while another pounds it, using the man to absorb the shock of straightening.

A GRATEFUL REFUGEE KID'S RECOLLECTIONS

(I have included these detailed observations of Deritend's process, because, for those of us who have never witnessed the extraordinary difficulty of casting such precise parts, it is a bit of an eye-opener. We witnessed another, equally time-consuming and complicated process at Deritend Vacuum Castings in Droitwich, which you may find of interest:)

At this plant Deritend makes high pressure and low-pressure turbine blades having controlled wall thicknesses. These blades are used in Rolls Royce engines and also for military applications. Deritend uses nickel and cobalt base super-alloys here with ceramic inserts and operates with a vacuum-melting furnace. Here, as in many other places visited, there is a need for a computer program to minimize gating, pattern assembly, mold framing and generally to program metal flow through an optimal runner system.

Only 30% of the metal poured is product-the remainder is framing and "waste" which has to be re-melted. Here too, a robot with extensive human-like wrist manipulation is used to place the mold into a slurry, at an angle, turning it, removing it, turning it 180 degrees to drain off the excess and then reversing. It repeats this same operation in generating the sand coating. The robot runs for 3 shifts.

The cast blades are given 15-hour high temperature solution treatment and then stored in a freezer to chill the warped castings before straightening. Castings are loaded at 3 PM one day, heat treated and quenched for the 8AM shift the next day so that the aluminum doesn't take a set or stress. Straightening by highly skilled people is done with 5, 10, 15 and 20 pound hammers. Quality control is very tight to assure high quality blades, especially for the high-pressure turbine blades. (quite an eye-opener, wot?)

What was also of interest at Cosworth, was, that they were able to recycle 99% of their Zircon sand from their molds. They used a thermal recycling process in a 1 ton/hour unit, with virtually no effluent.

JOHN LARRY BAER

Our "findings" in Sweden and Portugal

The next stop was Stockholm in order to visit *SAAB-Scania in Sodertalje, Sajo Maskin in Jonkoping* and the AB Volvo Car foundry in Goteborg, under the sponsorship of the *Foreningen Svenska Verktygsmaskintillverkare* (FVM) (Production Engineering Research Institute) (and you thought only the Germans had long words linked together.)

Going Cosworth one better, Volvo, using CAD with heavy emphasis on finite element analysis, was busy designing safer, stronger cars and planning to use 50 Kg (110 lb) magnesium in place of 200 Kg iron engine block, an aluminum steering column and a magnesium engine with iron or ceramic inserts, even for its trucks. Even in 1984 Volvo had a project to lighten car parts using carbon fibre for door arches and injection molding for roof, hood and outer panels. Their work included a 3-cylinder diesel with multi-fuel capability, a counter-rotating flywheel, already in use in city buses, and designs to recycle worn-out parts-lifetime conservation of fuel and materials, to include total recycling of all parts at the end of a car's life. (One little aside-when I mentioned that the workers in the castings finishing facility looked different from the other workers, I was told that these were "the niggers of Sweden"-workers from Finland.)

Saab-Scania had similar work going on to use composite materials, heretofore used only in aircraft. Their vehicle testing facility, a huge, multi-axis jolt and jumble table, sits in front of the plant so that all can see how well Saab tests its cars before releasing them to the public. It was also at Saab that we saw our first assembly team of some dozen young men in their late teens, led by a foreman in his twenty's. It was totally different from our assembly line process and, obviously, worked very well for them, producing cars with very few defects. The team's pride, like the need to maintain face by not passing forward a part less than perfect in Japan, promoted built-in quality-a concept later highlighted in Japan as "building quality in, rather than weeding poor quality out."

Findings at SAAB-Scania in Sodertalje, Sweden

According to Erich Schwabegger, the Foundry Plant Engineering Manager, the SAAB-Scania Iron foundry, rather than making their own patterns, buys them from Germany for the two sizes of 6 cylinder truck Diesel engines and the 3 sizes of cylinder heads as well as the gasoline engine they cast for Saab cars. They have a 4-station core manufacturing set up with automated handling.

SAAB built a new green sand facility in 1980 and (like Cosworth) manages to recycle 95-96% of their sand. The blending of their pattern sand is totally automatic. Absent any human operators, sand samples are taken automatically ever hour and sent to the lab.

SAAB's new molding line and continuous BUD machine were described in the 13 June 1983 issue of Giesserei. Their two grey iron cupola furnaces are out of doors under a shed roof, making the tending of these furnaces probably one of the least desirable jobs in the foundry, which, by the way is staffed about 80% with unskilled Finnish laborers.

For this process an ASEA robot moves the casting from the conveyor, turns and locates it in the press to shear off the flash. Compressed air is used to clean out the inside of the cylinder head castings.

SAAB uses a Modicon System computer to control the process mold line in making 1000 tons of castings per month with about 2 to 6% scrap. Their new conveyor system operates smoothly and quietly and facilitates what little human intervention is required in the core manufacture and lay-up. A Tellus Robot picks up and inverts the cope, puts it on the drag, which already has inserts placed in it by a woman who also inspects it. Another robot then turns and transfers the mold to the next conveyor. All in all a good automated layout, which makes 60 molds per hour.

Pouring, pattern and shakeout stations are all controlled with remote TV stations from an enclosed office. Even the timing and load of iron to pour are computer controlled to permit interchanging different items on the pour line even though the entire pouring process is totally automated.

Cleaning and deflashing of Diesel engine castings is still done by hand with chisels and power grinders in individual, soundproofed enclosures, because these castings need visual inspection that cannot yet be done by robots. SAAB was in the process of installing a new shot blast machine and automatic handler for cylinder blocks, which will permit their transfer from shake out to shot-blast without human intervention. SAAB also uses automatic slag skimming to obviate the need for this miserable job by a human operator.

It was rather interesting to note that the SAAB automotive assembly line teams, which were among the worlds first to use the cell concept, consisted 90% of young men between the ages of 18 and 28, all very dedicated and working very smoothly as a team in pleasant, relatively quiet surroundings.

Findings at Sajo Maskin in Varnamo, Sweden

Sajo-Maskin uses broken tools sensors, i.e. a station where a probe checks to see if the hole has been properly drilled. If it is, contact switches will advance the machine to permit a tap to move into place. If the probe indicates the hole is incomplete, a new drill is brought in to complete drilling the hole and the old, presumably broken one is marked for replacement.

Of interest was the fact that Sajo-Maskin, unlike some Spanish.machine tool builders and even Georg Fischer, does not fill any of their machine bases with concrete (for stability) but uses only welded iron bases. Generally they use cast aluminum and iron but also some fiber-reinforced plastic parts.

As one expects of Swedish products the quality control in the shop was very tight and the quality and finish of the machinery was excellent.

One of the problems cited was that in selling to the Soviets, the Russians are reluctant to pay in cash and prefer to barter, providing 1/3 of Sweden's oil and such other amenities as caviar and pelts, in exchange for machinery.

Of greater interest was the fact (even then,) that Sajo is planning to work with China to build a machine-tool factory in one of the free trade zones being created by the Chinese-and the Chinese pay in Dollars rather than rubels or furs. Trade with Roumania and Bulgaria was somewhat hampered by the amount of red tape involved, but the owner's son noted that Bulgaria is the country having a high reputation for building fine tools and equipment, replacing Czechoslovakia and its Skoda works who lost that reputation some 12 years ago.

Findings at Volvo Car Corp in Goteborg Sweden

Walking through the Volvo Car assembly line, like visiting a single GM assembly plant, is seeing only one very small part of the elephant. Nevertheless, one learns from everything one sees. Our host stressed Volvo's concept of autonomous teamwork, using lots of sub-assemblies of various modules and a short final assembly. Volvo uses CAD with emphasis on finite element analysis to design safer, stronger, not necessarily lighter, cars.

Volvo appeals to a different market than Citroen with its high tech car and engine and its extensive use of plastic parts; rather, they compete more with BMW and Mercedes Benz. Volvo makes extensive use of (and takes great pride in) in-house designed AGVs (automatic wire guided vehicles) for material transfer, rather than conveyors. Their plants are all highly automated and use hydraulic powered Cincinnati-Milacron welding robots in addition to dedicated automatic welding set ups.

Lars Bjerde, our host, noted that Volvo's truck sales to Iran account for 1/3 of Sweden's oil received in exchange. They sold 85,000 to Iran last year and 6000 to Iraq and have a truck assembly plant in Iran capable of turning out 1000 per year as well as an engine plant in Tabriz.

Close to Volvo's Auto Assembly plant outside Goteborg is their visitor center with its display of their LCP 2000 light component car. This project seems to have wide international support (*in some countries, like Japan and Korea*) and looks not only at lightening car parts, with magnesium wheels, engine block and chassis, aluminum brake drums, nuts & bolts, carbon fibre door arches, injection moulded roof, hood and outer panels, but also at its total lifetime energy consumption through drag reduction, counter rotating flywheel (already in use on city buses), 3-cylinder Diesel, advanced transmissions, multi-fuel capability and generally improved fuel consumption and even recycling of scrap automotive parts once they are worn out. (Compare this forward-looking attitude of the Swedes, to our American automakers' emphasis on profits 23 years later!)

All this is prefaced on the outside of the building by a programmable articulator, putting a huge dump truck through a torture test, comparable to that envisioned by TACOM (our US Army Tank & Automotive Command) in its acquisition of the Power and Inertia Simulator from Brown Boveri.

Other points noted about Volvo's auto plant are the very quiet and fully automated body pressing line, the 96 multiple-head robot welders working in synch with 120 people in 10 teams in their plant #2 and their insistence on, and thorough verification of, incoming materials quality to assure attainment of required specs and standards. This kind of QC of materials and subassemblies supplied to Volvo could certainly help American firms, including the Lima Army Tank Plant with its highly variable input of armor plate from different manufacturers.

The major points noted in Volvo's truck assembly plant, beside the much slower pace, are the trees growing in the center of the building around the quiet coffee area, the air lock doors to assure clean air and the temperature control, the very extensive, automated parts storage and retrieval system and the clever, small transfer cart with a skateboard on the back.

Numerous boxes were marked for shipment to the Zamyad Co. in Iran, as v/ere 2500 F12 Turso trucks. According to our guide, Iran like the USSR, which had 125 of the F12s on order, they want delivery in 2 weeks, but are very slow to pay. (This is a lot of space to devote to the findings about just one of the 40 plants we visited, but the lessons we should have learned (and implemented here in America) are enormous.)

Lessons from the International Foundry Congress in Lisbon

Leaving Goteborg *(Jöteborg)* with a short stop in *Köbenhaven* (Copenhagen) to walk through the famous amusement park and pass the less savory sex shops on the way to see the statue of the Little Mermaid, the next stop was *Lisboa* for the Lisbon International Foundry Congress held at the famous *Gulbenkian Foundation* that included a most interesting Robotics workshop. At that time Europe and Japan were far ahead of the U.S. in using robots to do the dirty, dull and dangerous jobs. Some of the observations at the Congress, were:

Maurice Grandpierre, *Administrateur, Centre de Recherches*, was of the opinion that the "new" cast irons will be cheaper and better than at present; that they will be fabricated under improved manufacturing controls and with new additives to improve their strength and toughness, thus permitting thinner wall sections and also improved machinability characteristics. He believes that the industry will have to look at each component for its unique requirements and characteristics-engine blocks, cam and crank shafts, etc.

Allen G. Fuller, BCIRA, Alvecaurch, Birmingham, England, noted that robots used for trimming castings can readily be adjusted to remove light or heavy flash or even a stub, but that most current robots need to be operated in a very clean atmosphere. (This is not at all the case observed for the trimming and grinding robots operated by Citroen in their foundry.) He was also of the opinion that it is better to program the robot to pick up the casting to be trimmed, rather than have it manipulate the tool. (Again, Citroen is doing just the opposite, as also advocated by Renault, and doing it quite successfuly.) (interesting– each expert has his own opinions and preferences.)

Wolfgang Sturz, Fraunhofer Institute, Stuttgart, W. Germany, emphasized the need to use electronic feedback to measure tool-wear during machining, to use tactile sensors to check burr size, to measure cutting and clamping forces and tolerances, as well as robot power consumption. He felt that robots may be too slow to react where tight tolerances are required. (this is contrary to what we observed at MBB *Messerschmidt-Bölkow-Blohm.*) (see page 95)

After a brief visit to the 100 year old Cometna steel mill in Porto, Portugal, it was back, via Madrid and a weekend visit to *Jerez and Seville*, to France, on our way North to the *GIFA International Foundry Trade Fair in Düsseldorf, Germany.*

Renault Machines Outils Fonderies du Poitou

ZI d'Ingrandes sur Vienne in Dange Saint Remain, France (train station Chatellerault); *Maurice Deschamps*, Directeur Industriel

Monsieur Deschamps felt that two significant thrusts in the foundry in coming years will be a search for material economies and for new crucibles. His foundry manufactures aluminum cylinder heads using low pressure die casting and side gating, which, with 2 castings per pour gives them an 85% yield (compared to the poured aluminum cylinder head castings at Citroen which have only a 50% yield.) (Just a little comparison of two major French auto makers.)

Renault fonderie had a very high degree of automation, rapid machine operation on their sand mold baking machines, extensive conveyor transport of patterns and castings in a relatively new, modern plant. However, they still used quite a bit of manual core handling, pattern cleaning, blowing off excess sand and spraying the molds.

Renault used extensive sound baffling between core making machines and between manual grinding, deburring and cleaning operations on the iron engine blocks. It was interesting to note that with the manual application of rapid glue to assemble their 3 piece cores, which are carefully hand-trimmed, these are much sturdier than similar pieces observed in the US and elsewhere. They could be struck sharply on the side of a metal container without breaking.

(This seems like a lot of space to devote to modest technology, but, as noted before, it makes for an interesting comparison among two major players in the auto manufacturing field.) And why all the detail about their location? Thereby hangs another tale. After leaving the foundry, we drove on *green* (i.e. secondary) roads through the countryside, till we came to a little country restaurant that had been recommended to us by *M. Deschamps*. We drove into their tree-lined courtyard, trying not to hit the geese who greeted us raucously, and, as we parked, observed a man come up from the brook at the foot of the property with a bunch of freshly caught fish. Needless to say, that was our delectable lunch, prepared to perfection. Happily, it was one of many instances during our many travels through Europe, that we were treated to such fresh, tasty fare.

Findings at *Citroen Usines des Ayvelles, Charleville-Mezieres*

One of the memorable moments from our visit at Citroen, was when I mentioned the concern of my sponsor about minimizing flash on castings. Our escort, Monsier LeBrave, the *Directeur des Methodes Generales,* said: "But, Monsieur–why do you make it with ze flash?" (as in baking a cake) he said that: "you must keep clean the surfaces of the cope and drag (the top and bottom of the mold) and seal them well with pressure, and then you will not have ze flash." (such simple advice, but not always attainable when dealing with high pressures.)

The GIFA* '84 International Foundry Trade Fair in Germany

I had visited the Chicago Foundry Expo in the massive exhibition hall, but I was not prepared for the huge Expo Center that awaited us in *Düsseldorf.* Nor had we reckoned with the huge turnout of visitors, who had booked every spare hotel room in town. Fortunately, my cousin Erich, in Bad Herrenalb, through some of his connections as a hotelier, was able to find a charming, well appointed room for us in a private home, within easy walking distance of the fair. We had our own private bath, private entrance and a typical German *Frühstück* every morning before we set out for the Expo. Best of all, our hosts told us about their favorite local *Weinstube*, which we visited almost every night to delight in their well served, savory suppers.

We even found time for a stroll through downtown, to see the black swans in a small lake in the middle of town and to attend a splendid performance of *Lucia di Lammermoor* with the then still young, unknown June Anderson.

But our purpose was to learn what we could from the hundreds of GIFA (**Giesserei Industrie Fabrikation Ausstellung*) exhibitors scattered among the nine halls in the exhibit area. Thank goodness I had my trusty aide by my side. That way, on our 3rd day, she was able to tell me the name of the East Russian, whose memorable technology exhibit we had seen on our first day in *Halle 1*. What he was showing was a Russian made ceramic filter cloth, through which they poured the molten metal, to filter out any air bubbles or detritus, thus yielding a purer melt. Interestingly, we later found that the Israelis were using the same technology in fabricating the armor for their *Arkava* tank-but we couldn't interest GMCFD or any other American foundry to use it. NIH at work! (not invented here!)

Findings at the Daimler-Benz *Leichmetalgiesserei* (light-metal foundry)

Naturally, a trip to Germany would not have been complete without a visit to the *Eisen und Aluminium Giesserei* (iron & aluminum foundry) *at Daimler-Benz in Stuttgart*. Here, as in virtually every foundry we visited, the thrust was for a clean work area to produce clean castings-another lesson not easily transmitted to the sponsor. Whereas at GMCFD you couldn't see clearly from one machine to the next, at Daimler-Benz, they prided themselves that their 1.3 million cubic meter per hour air recycling system permitted you to see from one end of their football field long foundry to the other. And with mechanical floor sweepers constantly making passes through the plant, you could have eaten off the floor.

As an interesting aside, as our guide, the *Leichtmetallgiesserei Leiter, Dr. Mollenkott*, walked us through the plant, he spotted a beer can on the floor (yes, German workers can have a beer on their break) he stooped to pick it up, and with a no-no wag of his finger, handed it to the man at the nearest machine. I recall also that when *Diplom Ingenieur Detering, the Direktor of the Giesserei* showed us the machine casting area, we witnessed what looked like a perfectly fine engine casting, being scrapped. When I asked about it, the Direktor showed that the casting had a fine crack in it, one which I thought could have been easily "buttered." But he said, *"Nein, Nein-*if the car was ever in an accident and such a defect were spotted, even though the engine was perfectly safe, people would question what else had been 'fixed' and would lose their trust in the Mercedes strive for perfection."

By the way, 70% of the 200,000 tons of sand used by Daimler Benz daily, is regenerated. Naturally, they would like to be able to regenerate 100%, which they can do if they switch to cold box but the sour hot-box and croning sands do not permit this at present. DB uses the first European 100% hydraulic core-shooting machine. All operations are completely controlled from a central control booth using ASEA controls, including quality control data transmitted from the laboratory.

Also of interest, Daimler-Benz calculates wages with special factors added for exposure to heat or noise, lifting required, etc. (something we might consider in the US as well.)

Findings at *Bayerische Motor Werke, München*

Somewhat ahead of our American thrust at "Concurrent Engineering," BMW stressed close working relationship between their designers and manufacturing engineers so that future concepts can be anticipated in the factory.

BMW's sand knock-out is totally below ground and totally automated, but their inspection is very much above ground and of great concern to them. All wheels and drive housings are fluoroscoped

100%, as well as receiving external UV inspection of coated and washed parts. Cylinder heads, which leak in tests, are dipped in waterglass and cured 3 days, then re-pressure tested one at a time under water to assure that they pass the leak test. With all this, BMW limits scrap parts to less than 3%! Not bad.

The BMW foundry, whose, small, clean, orderly layout can be comfortably viewed from a glass enclosed, overhead walkway, is unusually clean and quiet, even on the work level.

Another side trip took us to the *Giesserei of Messerschmidt-Bülkow-Blohm* in Augsburg, not far from *München*. Here, of course, the thrust was aircraft engines, 90% of whose Titanium castings had to be milled away to achieve the lightness and strength they needed. Here we were shown another noteworthy technology. Each milling head was fitted with a matched set of four milling cutters. When one cutter broke or became dull, the whole set was removed and replaced, and each cutter sharpened, measured and matched up to form a new set, which was then numbered and stored in a drawer until it was needed. This was a fine example of computer control, in which MBB was well ahead of its time.

Findings at *Urdan & Vulcan* Steel and Iron Foundries, *Natanya & Haifa*, Israel

Dov Nardimon, the General Manager, and Dr. Yitzhak Sharir, Technical & Quality Control Manager at Urdan, both felt that much work in the foundry, even today, is more art than science and that the foundry industry generally is chary of innovation. They also suggested that it might be of interest to look at foundries in Arab countries, who have had direct help from Russia, especially Egypt and India who, like Iran, have had Soviet plants installed. Note that Israel had adopted the use of Russian ceramic filter cloths, that I had noted before, to give them clean castings for their tanks, while our (we can do it better) foundries, did not.

At the Vulcan Steel Foundry in Haifa Bay, we learned that, unlike many of the firms we had visited so far, they had no *Gastarbeiter*. Not a single Arab worker in the place. We also had enlightening visits at the Israeli Institute of Metals at their Ministry of Defense and at the Technion University Casting Lab in Haifa. The main interest expressed in both places, was the American work on using new types of cast maraging steel with very high tensile strength, using ceramic parts for engines, direct solidification of turbine blades for aircraft engines, and American super-alloys for use in their military hardware. So there were some areas were U.S. technology was ahead of the Soviets. So now it was on to Asia (via Paris,) there being no direct flights to Hong Kong from Israel (at that time.)

Observations in Hong Kong

We quickly learned that Hong Kong foundries are generally small and limited to the manufacture of small items; thus there are no machine tool manufacturers of any size to be found in Hong Kong. Any large castings required are imported from Taiwan (or China if the demand for quality is not great.) For more accurate, but more costly castings they turn to Japan or Korea.

At Chen Hsong Machinery in Tai Po we found that they build 500-ton capacity plastic extrusion machines and plan to build 1000 ton machines in the fall. All their machines have numerical controls for temperature, velocity and pressure controls and read-outs. The rheostats of former days have been replaced since five years with these very precise controls. Their major castings come from Taiwan and, where better accuracy is demanded, from Japan, at a correspondingly higher cost. They are looking to China as a future source of castings! (over 20 years ago!)

Their international business manager, Mr. Lee noted that Hong Kong workers are not as industrious, dedicated or loyal as those in Japan and that their skill level is lower than that found in the US. He feels that Hong Kong, while a desirable climate to build small units, does not compare to Taiwan, Korea or Japan in terms of quality control or quality level.

Findings at the China External Trade Development Council, *Taipei, Taiwan*

Well–first of all, we almost didn't make it to our appointments! We left our splendid hotel in *Hong Kong* on a short flight to *Taipei*, relying on the guidebooks that we could get the entry visa we needed at the airport. Lo and behold, upon arrival in Taipei, we were advised that since America had decided to "recognize" (mainland) China, we had to have visas prior to entry. Despite our pleas about having hotel reservations and appointments next day, and since we had no embassy in *Taipei*, only a consulate, and it was closed, we were taken under our arms and put on a plane back to *Hong Kong*!

The kindly hotel manager was, naturally surprised to see us back, but when we told him our story, he said not to worry. "Come down in the morning with your passports and $ 10 and we'll send a boy over to the embassy to get your visas." Without further ado he assigned us to a huge suite overlooking *Kowloon* Harbor, and, after a delicious dinner, we went to bed content.

Next morning, sure enough, while the hotel "boy" went to get us our visas, the manager directed us to visit the Stanley Market to spend the morning. It was interesting, of course, but we had done our shopping days earlier and it was so hot that my good wife said: "the next place that has air conditioning, I'm going in. I don't care if it's a brothel."

Happily we quickly came upon an air-cooled art gallery, which she entered and within a few seconds came running out, saying: "I found it." The "it" turned out to be a wonderful bronze casting of the *Han Horse*, which we had seen at our National Gallery during an exhibit of Chinese art. Despite its considerable weight, I was happy to carry it from Hong Kong to Taiwan to Korea to Japan and home–where it has adorned our dining room ever since.

The next day we sojourned once more to Kai Tak airport for our flight to Taipei. This time, armed with a proper visa, they let us in to the country. A taxi took us to the Grand Hotel, which had previously been Madame *Chiang Kai Chek's* palace–or so we were told. And so it was on to our meeting with the director of the China External Trade Development Council.

That gentleman, *Chyou Wen Yueh*, advised that there is very little activity in Taiwan that is related to the auto industry and that large castings are made only for ships.

He also felt that there was virtually no new technology being developed in Taiwan, but that they were mostly copying US and European designs, using European technology and looking for a quick way to make money. (Remember–this was 1984.) Mr. *Yueh* noted that most Taiwan factories employ less than 20 people and make simple products. The castings they make are "cheap" and not of high quality according to Mr. Yueh. But here, as in all the other plants we visited, cleanliness was essential–even if it meant sweeping floors with a straw broom.

Our next stop was the King Kong Iron Works. KKIW works in the middle range of quality to turn out machines at reasonable, competitive prices, but they are working toward higher quality, using a 4-axis CNC lathe. (Lio Ho, whom we visited the next day, noted that they had a high rate of rejects in material ordered from King Kong.)

The Yang Iron Works in Taichung, is actually a machine tool manufacturer. They compete directly with Korea and Japan in machine tool sales and their CNC Horizontal Machining Center with Fanuc controls have been sold to Ford-Taiwan. Yang is a generally clean and well-run factory in buildings largely open to the atmosphere to let air flow through.

The Metal Industry Development Center in Taichung, according to its director, makes good aluminum die castings for the thousands of motorcycles sold in Taiwan (they outnumber cars about 10 to 1) but he felt that most of their conventional castings are not very good (quite an admission to make to a foreign visitor.)

The Lio Ho Machine Works in Chung Li, formerly well known for its machine tools, had at the point in time, stopped making machine tools and was entirely devoted to making castings at about 350 tons/mo, using new automatic molding machines. These castings will be for Ford, Mitsubishi and China Motor Co. They also made fender and hood stampings for GM-US (supposedly at half the cost of US made parts), engine support castings for Ford-US, castings and stampings for Westinghouse and for IBM computer frames.

Lio Ho used the lost foam process back then to make fender mold castings with great precision and also produce very high precision PC frames for IBM. We also observed that their aluminum wheels, even as cast, had an excellent finish and their automated riser ring removal quickly readies them for the 100% X-ray inspection & leak test which they receive.

In sharp contrast to King Kong Iron Works, where all the parts were stored in heaps on the ground, the finished product at Lio Ho, and even work in process, was neatly stored in appropriate racks and containers. Lio Ho gave us the impression of being a first class operation.

Findings in Seoul, Korea

Surprisingly, back then in 1984, we found relatively little of interest in visiting several Korean foundries. At the Korea Trade Promotion Corp. we learned that the government was supporting new casting technology for auto parts, including stripper technology developed in Japan.

Shin Han cast iron foundry, about 30 minutes out of Seoul, casts about 950 tons/month of small malleable cast iron fittings, of which 60% were exported to Iran and the U.S. We found Shin Han to be the dirtiest establishment we had seen on our entire trip. The other foundry we visited in Yongsan-ku was not much better.

And then there's Japan!

We had expected the German and Japanese foundries to offer us the newest technology and best levels of quality management–and we were not disappointed. The visits to the SME/ Mechatronix/Autofact

Show and the plant tours of Kubota and Komatsu in Osaka set the tone for expertise.

The Mechatronix/SME Autofact 8 Expo was held at the Minuto Fairgrounds in Osaka, Japan. The Autofact portion showed several joint US-Japanese ventures in CAD/CAM/CAE hardware and software, nothing that could not be found at any US show and without the handicap of having everything in Japanese and virtually no one who spoke English. (Remember, this was 1984. Now most foreign engineers speak good English.) Mechatronix presented a relatively small display of robotics including Komatsu Visual Sensor, welding and Bolt Tightening Robots; and a stud-driver robot which will pick up a bolt from a rack, place it and then tighten it to the bolts yield point, which it senses through torque and rotation angle. This show after the elaborate and informative GIFA exhibits, was a great disappointment. OK–on to *Kubota* in Naniwa-ku, where Kei Takada, Chief Executive of the Materials Consolidation Division was our personal, and most informative, escort.

If the Osaka fair was a disappointment, the *Kubota* visit and the others described below, more than made up for it in providing useful information. Mr. *Takada* told us that the foundry represents only 10% of Kubota's work. He foresees no drastic changes over the next 3 to 5 years, only some gradual process changes such as maybe making iron castings with thinner walls and some increased use of filament reinforced metal. He considers ceramics too expensive and not economically applicable at present for auto engines, even if the cooling system can be eliminated. In 10 years, (i.e. 1994) maybe with Diesel engines, which he expects to be in greater use then, there may be more use of ceramic inserts and components. He foresees a need for casting closer to net shape (maybe with the lost foam process) and greater use of robots for finishing operations performed manually.

During our visit to the *Okajima* plant we saw some of the large 1-piece precision die-castings they have made such as a 1.5-ton turbine blade and a convertible pitch propeller. Mr. Takada noted that they and Japanese car makers will be following closely the results of GM's Saturn project for making lighter cars which, like Volvo's LCP 2000, would require lighter engines and other car parts.

At the *Okajima* plant we observed ASEA robots handling parts and transferring them to a machining center in an enclosed, soundproof booth for automatic flash removal and deburring. There seemed to be a large number of hollow core patterns, which Kubota and other Japanese foundries like, because they reduce the amount of sand used, and sand, in Japan, is more expensive than in the US-see page 108.

Of interest also was a new style transmission housing they were casting for Honda, mostly with shell mold cores. They also used *Dainichi Kiko* robots to place transmission housings and 4-cylinder engine blocks into fully automatic inspection and dimensional checking machines.

Komemushi-san, the director at *Naniwa Foundry Machinery Products* in *Yodogawa-ku*, Osaka pointed with pride to their quick die change capability and greater clamping power, which results in less flash with their equipment, though it does not result in completely flashless castings. However, parts from their equipment are virtually without parting lines and have very little flash. The more accurate the core, the better the casting and the less the finishing needed.

Mr. *Komemushi* noted that Japanese foundries use more hollow cores to reduce core costs, as sand is very expensive, while the US can afford to use solid cores. He also felt that US foundries might be well advised to use Naniwa machines to cut their 50% scrap rate in core production.

At *Komatsu*'s main office in *Hirakata* City, Osaka, we were welcomed by K. *Furukawa*, the Senior Manager for Public Relations *(see photo.)* After a brief overview about this major corporation, he gave us a tour of their Construction Equipment Department in Akasaka before turning us over to *Takafumi Imada*, Robot Engineer in their Machine Tool Division in *Komatsu City, Ishikawa-ku.*

One of the interesting points made by Mr. Imada at the Mechatronix fair, was that *Komatsu* makes 60,000 engines per year at their *Oyama* plant, north of Tokyo, using a high degree of automation and robots with vision sensors. We observed how these robots inspect multi-layer welding on their huge, heavy-walled construction equipment, very similar to our Abrams tank. In this way, if a void is detected, it can be repaired immediately. By contrast, in welding our tanks, we do all 15 or 16 weld layers before inspecting for voids! That way, if a void or inclusion is detected by the X-ray, we have to grind down to the suspected void, repair it, and then re-weld all the layers over it–and then re-inspect! We could learn something from Komatsu.

At the Osaka plant, Komatsu makes tracked construction equipment and other heavy off-road vehicles, using an integrated production system. Bolts and nuts on their 80-ton bulldozer are tightened automatically with a multi-head bolt tightener and air bearings are used to move the behemoth around the assembly floor. Rather than putting tractors through actual rollover tests, the way we test jeeps at APG, they accomplish the same safety testing by simulation.

Komatsu has also made automatic pipe layers, which were sold to USSR. Their tractor and bulldozer assembly moves along at a constant 190 mm/min on wire guided transport vehicles, with 13 different models being handled on their assembly line, which is highly automated and very clean. They export equipment to the US, USSR, China, Iran, Saudi Arabia, etc. On large stretches of welding they use a mechanized set-up, but smaller sections are welded manually.

Their green sand set-up produces 80% good molds; they have an excellent air cleaning system, which confines smoke to the area atop the furnace where it is generated. Track shoes are cast in a very clean state with 6 mm thick sections, with only 1% scrap parts. Four torches are used to automatically cut off gates, which represent about 30% of the total casting weight. Komatsu people believe that they are well ahead of John Deere, IH and Caterpillar in that they assemble bulldozers, pipe-layers and other comparably large and complex equipment all on the same line.

They noted that the demand for hydraulic excavators in Japan is about five times that of the US. They sold 2700 pieces to the US in 1983 and Japan's demand is for 25,000 per year. They expect to sell 4000 units to the US in 1984. Over 90% of their bulldozers and 40% of their excavators are exported. Komatsu is looking for further plant locations abroad, beside those they have in Brazil, Mexico & Indonesia and may set up a plant in the US and/or Europe. Before they do they want to look at the results of Honda's car assembly plant in Ohio. They have already sold a 120-ton capacity dump truck to the US and maintain repair and maintenance capability in the US. Some parts for their equipment are already made in the US.

(One might note in passing, that Komatsu could, without doubt, convert their line to the manufacture of tanks and tracked combat vehicles in very short order, and do so very effectively!) (Also, their doubts about the US worker should be assuaged quickly when they read about the Nissan truck experience at their Tennessee plant- Washington POST of 12 August 1984.)

Shoji Matsumoto, the Senior Vice President at *Ryobi Limited in Hiroshima-Ken* sees an increasing level of quality control in the manufacture of castings in the future but no major changes in the types of castings or the equipment or method to make them. He also sees an increasing drive to decrease the porosity in castings by controlling the entrained gases. (Vulcan foundries are already looking at this aspect, using the previously mentioned Russian technology.)

With conventional castings they estimate 30 cc gases are entrapped per 100 grams of casting; by using a gas free process this can be reduced to 15 and if a so-called pore-free casting can be created in an oxygen atmosphere, they would expect to approximate zero cc per 100 gm. However, like most foundries (except Vulcan) Ryobi at present checks only the composition of the melt, not the entrapped gas content. (Maybe they should consider using the Russian ceramic filter cloth.)

In summary, Mr. Matsumoto foresees changing from sand and low pressure or permanent mold casting to die casting to achieve increased casting strength, soundness and quality level. His greatest interest is in eliminating porosity in castings and the procedures needed in the future to accomplish this.

Ryobi uses computer numerical control (CNC) of piston speed to check piston wear and to compensate for this wear by adjusting both piston speed and pressure. Like some other foundries visited (see above) Ryobi feeds molten metal slowly at high pressure as a squeeze casting technique to reduce porosity. Toyota also uses high-pressure squeeze casting to make auto wheels.

To achieve castings of improved hardness and wear resistance, they have tested fiber reinforced aluminum and are experimenting with silicon carbide and silicon nitride powder, which are easier to mix and obviate the need to worry about fiber orientation.

Most auto parts made by Ryobi require only trimming and a little hand filing by the die cast machine operator. There is lots of automatic extraction handling and trimming, designed and manufactured by Ryobi in making such parts as aluminum oil pans for the Cummins Diesel engine. Automated manipulators are used to remove parts from the die-casting machine and move them directly to an automatic trimmer. After the part is trimmed it is lifted and moved forward while the scrap is brushed off in the opposite direction. The parts then go to tilt table where they are transferred automatically to a basket. Pistons are trimmed, gaged and tested 100% by hand. As a result Ryobi achieves less than 1% rejection rate for transmission cases and less than 2% are rejected by the auto manufacture when he checks them for porosity.

Yoshiaki Kanno, director of the *Honda Motor* Overseas Public Relations Department in *Jinguma*e, Tokyo, noted that Honda is looking at ceramic engine parts for the future, but that implementation is a long time away, unlike plastic parts, which are already used extensively in the Honda to reduce weight while having a durability often greater than steel body parts. *(Now, 23 years later we find extensive use of plastics in both Honda and Toyota cars.)*

Honda uses computer-designed engine casting molds for their cast iron blocks and aluminum heads. Honda also plans some reconfiguration to permit more automatic bolt tightening. He also noted (ref. Mr. Matsumoto's concerns) that Honda workers in Ohio work just like Japanese workers in Japan; they are responsible for their machine and the product it turns out; they rotate jobs every two years and they have no Union.

At their *Saitama* plant, Mr. Kanno noted, that they have 19 general-purpose welding machines, which can make 500 welds simultaneously and which have heads, which can be changed in 10 minutes.

The main thing we learned at *Showa Denko* in *Minato-ku*, was that many parts with limited apertures are difficult for them to spray due to the large size of their plasma spray gun; thus it cannot be used to get into small spaces to provide coatings of uniform density and coating thickness. They feel that, to be effective, they need to get a 1 mm thick coating and have only been able to achieve 0.5mm Zr oxide coating so far, and this has not always held up as well as they would have liked. Their coatings thicker than 0.5mm tend to crack at present. By comparison, the Cummins diesel engine being tested by TACOM, with an 0.6 mm coating has been running successfully in a 5-ton truck for over 10,000 miles.

Dr. *Masanori Moritani*, Senior Researcher, of the Nomura Institute in *Chuo-ku Nihonbashi*, is the author of an excellent book on "Japanese Technology." He foresees a "fine ceramics" boom in Japan, especially for use in electronics, but not in automobiles. He sees no interest from the auto design engineer who sees ceramic coatings of parts exposed to high heat as providing small improvements in fuel consumption, maybe 10 to 15%, possibly useful in diesel engines but not in gasoline engines where they would be costly to introduce. He does see a rapid growth in electronic packages for automobiles, including ceramic condensers.

As for the casting industry, Dr. Moritani feels that the Japanese foundry industry will introduce and implement fully automatic controls very rapidly due to its dirty nature. He sees only graduates from the middle of graduating engineering classes going into fields like the foundry industry, not the top graduates who opt for more exciting fields. He does see Japan leading the thrust for quality castings, citing as only one example Ebihara Pump Manufacturer's unhappiness with the quality of the Korean castings they received and their poor delivery times. (Nevertheless, I see exceptions to this in companies like Lio Ho who have demonstrated a major concern with quality and timely deliveries.)

Dr. Moritani sees the already inexpensive NC (numerically controlled) systems getting cheaper and still smaller, especially in Japan, and thus being applied to progressively smaller machine tools, by ever smaller manufacturer. He also sees plastics as having a sizable impact on the casting industry, especially with fiber and powder reinforced plastics competing with more conventional, heavier and costlier materials.

***He noted that the current slogan of the Japanese Electronics Industry is: *Kei Haku Tan Sho*, which means "Light, Thin, Short, Small."

He sees high tensile steel plates getting cheaper in Japan and augmented by light weight, new technology, dual phase steel for possible application in automobiles, which could use welded, light weight plates to decrease car weight. He feels that Japan is very good in applying new electronic technology to iron and steel with bright young engineers to design the new electronic controls. Processes that used to take 7 to10 days can now be accomplished in a few hours.

Dr. Moritani sees Japan's trade imbalance vis-a-vis the U.S. getting worse and he believes that Japan is basically afraid of America's basic strength, but also fears growing competition from Korea, Taiwan and Hong Kong. He believes that America is strong in matters affecting security and national prestige (military prowess), but also in its frontier technology of significant basic research, its welfare technology and its business technology. He sees Japan strong in the business technology, especially in the mass production of trade goods.

Thus he sees a need for Japan to give more technology to the US as a quid pro quo for American technology implemented successfully in Japan, but he feels that Japan is afraid of the boomerang effect and is thus reluctant even in its technology transfer to Korea or other Asian neighbors. He believes that Japan, America and Europe must assume more of a world wide view in terms of its technology and markets of the future.

(How come a Japanese taxi driver can automatically open and close the left rear door of his cab to let passengers in or close the door after them, but in some crime-prone American cities this feature is not available to security conscious cabbies?)

You might well ask why, 23 years after the fact, I have included all of these prognostications by our eminent Japanese author. I did so, as you will also surely observe, because virtually everything he told us, has come to pass since then! But even this eventful interview was topped at our last stop in Japan–Fanuc, Ltd.

At *Fanuc Ltd.* (formerly *Fujitsu Fanuc*), at the foot of Mount Fuji or *Fujiyama* we were greeted by *Takeshi Kawamoto*, the Administrative Manager of the Fuji Complex, and *Masahide Iwai*, Chief of Public Relations in Corporate Planning.

In describing Fanuc's factory it is easy to slip into superlatives. It is also a good place to visit as the last stop on ones itinerary because all of the petty annoyances encountered during a 10-week trek around the globe, suddenly pale to insignificance when one visits what has properly been called by many, "the factory of the future." Unlike many companies around the world, they do not make small, incremental moves-at least they certainly seem to move in seven league boots.

Fanuc moved their robot manufacturing facility to the Fuji site in Sept.1980 and it was at the urging of the late Jack Williams, and thanks to his help, that we were able to see it. In Sept. 1982 they moved their control motor manufacturing to Fuji; in April 1984 their EDM machinery plant and in Sept. 1984 they expect to move their Hino headquarters and NC Equipment and control manufacturing to Fuji.

Fanuc now manufacture 350 6-axis robots per month with 70 people at Fuji, plus 15,000 control motors per month with 60 people and with 120 3-axis robots; they manufacture and assemble 110 EDM (electrical discharge machines) wire cut machines per month with 30 people, although they could build as many as 250 such machines if demand warranted it; and in the new plant they plan to build 8000 NC machine controls per month. They also plan to build a new plastic injection molding plant at Fuji in 1984.

One of Fanuc's Mitsui Seiki machining centers had 5 different parts on the pallet feeding the machine, delivered automatically by a computer controlled, wire guided cart, which unloads its pallet at the designated machine and then returns to its pick-up station. The machines operate untended at night, but do not perform any fine drilling at night since there is too much risk in breakage incidence of small drills and taps. During the day one worker supervises 5 such machines. All robots produced by Fanuc are put through 100 hour programmed, computer-controlled manipulation tests. (Again–why all these details in a *mémoire* about my life as a refugee kid, turned engineer? It's because we were so impressed.) When we were escorted to the viewing platform, overlooking the production floor, our host turned on the lights, and my good wife mused why there were no workers on the floor, tending the machines? Were they all out to lunch? It was only when we saw the AGVs (automated, guided vehicles) in operation, that we realized this huge factory floor was operated without human intervention, and thus could well function without lights.)

In the manufacture of control motors 1 operator tends 10 machines and, as described above, wire guided carts deliver raw materials to each machine, as needed, as for example, 48 housings in stacks of 4 on a pallet with a 3 x 4 array. Some of the machines are equipped for automatic, in-process gaging and sensing broken tools, but they have no wear sensors for cutting tools. However, machine tools built in the future may have tool wear sensing built in as a desirable feature. The condition of each machine can be monitored from an overhead control room with 3 TV monitors. Unlike MBB, Augsburg, which had attained 40-50% process time on their milling machines in 1982 and hoped to attain 70% machine utilization with their CIAM (computer integrated and automated manufacturing) program, Fanuc already achieves 62% machine utilization over 24 hours.

Fanuc's computer numerical controls, such as their "System 9" series can simultaneously coordinate up to 15 axis operations according to Mr. Kawamoto. Fanuc's 330 D is the world's first three-dimensional tracer control incorporating a computer for milling and die-sinking machines. Their CAD/CAM systems include a super-high speed 3-dimensional profiler. Fanuc merchandises 7 special purpose robots, 7 assembly robots and 5 machining robots with arc welding, distance and visual sensors, and their factory programming system permits making 450 different types of parts in 30 machining cells in lots ranging from 5 to 20. To walk into their motor or machine manufacturing or assembly shops is a bit unnerving in that, as noted above, there seems to be no activity. 35% of Fanuc's staff is employed in R&D, including CAD (computer aided design.) On the shop floor now, some workers will tend as many as 15 NC machines. The difference between Japan and most other countries is that when new robots are brought in, to replace workers, these workers are used in their areas of competence, because improved productivity has increased demand for Fanuc's machines so that there is still work for everyone to do. (This was also noted by Madame DuPont-Gattelmand in our meeting with her at Renault Machines Utils.)

Orders for CNC (computer numerically controlled) control units now exceed 3000 per month and they are now making 4000 per month just to try to catch up with back-orders. They do this without losing sight of their slogan, prominently displayed throughout the plant- *Reliability Up-Costs Down*. Their target is for a failure or reject rate of less than 2 problems per 100 month or 0.02, but they are presently achieving 0.03. As to vision systems for robots, Fanuc feels that since vision system costs are much greater than robot costs they will have limited use until prices can be brought down.

As to personnel practices, Fanuc has no job descriptions for their factory employees, who are all technical high school graduates and have been taught to program or run any machine on the floor. After training at Fanuc's school he can become a maintenance engineer or shift easily from machine operation to programming or sales, service or maintenance. In other words, he is expected to know not only how each piece of equipment is made but also how it works and how to fix his machine or the system being built. Workers start at 5000 Yen per day (about $20.) As a worker's productivity and his contribution to the plant grow, so does his salary. Support staff is slightly larger than operating staff and a small administrative staff reports directly to Dr. Inaba, but each department is relatively autonomous, not under the direct control of the director.

According to both Dr. Inaba and Dr. Moritani (Nomura Institute), the US is ahead in technology, particularly in robotic vision and tactile sensors (this quickly became apparent to Fanuc in their joint venture with GM.) However, Japan and the rest of the world are only too happy to use US technology and to go one step further by adding tight QC when implementing this new technology. Both men felt that on very large mass production we must watch out for India, Egypt, Taiwan, Korea and other developing nations (no mention of China then.) They may be crude and their work areas dirty, but they are effective and turn out a product, which is adequate for the user. (This same comment was also made to us by several different people at the GIFA, including a Turkish foundry manager, the head of the *Helwan* Iron Auto Engine Casting Foundries in Egypt, and many of the machine suppliers to whom we spoke.)

Japanese parts made for the civilian market are already up to US military specs and the reliability of their integrated circuits is 10:1 compared to US IC's. If there is a single piece of advice which came out of this trip it would be to take QC out of the QC department and put it back in the workers head. In Japan, the responsibility for bad parts is shared by workers and foremen, and even top management.

Thus the worker actually responsible for turning out a defective part, brings shame on everyone around him-something they all go to great pains to avoid.

That this attitude is not strictly Japanese has been born out by American workers at Japanese owned companies all over the US making cars, trucks and Video recording tape cassettes. This attitude, however must be instilled by top management and top management can only do this when they feel their own job is secure and their performance judged by a board of directors over the long pull-3 to 5 years, not 6 months to a year.

As long as we allow stockholders and top management to demand quick return on investment, countries like Japan and the emerging nations will beat us at our own game every time, taking business away from our factories with their lower wages, higher productivity, and high quality production, and America will become a service nation, providing technological "stud service" to the rest of the world.

OK–one more anecdote that tells a lot about Japanese people: We were on the *Shinkansen*, their "bullet train" on the way back to Tokyo. In those days, as maybe still, everyone on the train smoked, and I started to cough. Moments later, a well-dressed Japanese gentleman came over to ask if the smoke was bothering me? When I told him that it was rather bad for my chronic bronchitis, he bowed and excused himself. He quickly went to people in the nearby seats, and within moments we were in a "smoke-free" area. I thanked everyone profusely, with a *domo arigato gosaimas,* and we made our way happily smiling all the way back to Tokyo.

There is actually one more worth telling. We were sitting in the *Narita* airport in Tokyo, having checked us, and our luggage in, and waiting for our plane to be called. Instead, there was an announcement that they were over-booked, and anyone willing to give up their seat and take the next flight in two hours, would get $ 300. Of course, that bonus was too good to pass up, so I ran up with our two tickets, got $ 600 in cash (ah, for the good old days) and instead of flying via LAX, our flight, two hours later, took us to Minneapolis to make a connection. So I figured I'd better call our son and tell him that we would be late in getting to DC. He was blown away by our call from the US, because he thought we were coming in the next day–he had forgotten about our gaining a day, crossing the international dateline.

Whew! Reading all this, I feel like the joke tellers Edward Ford, Harry Hershfield or Joe Laurie. Jr. on "Can You Top This?" back in the 1940s. But, this was only the wonderful start of a 20-year second career. Once our report, with my findings and Dr. Cetron's prognostications, was finished, I delivered it to the GM-CFD manager, who had given us the challenge. He took me through part of the foundry I had not seen before and I noted all the trash–coke cans, coffee cups, cigarette packs, etc, in the wells around a casting pedestal. My host agreed that this was bad and they had hired a new VP, whose job it was to clean up the facility. I told him that I thought this was great. But when I asked how long the VP had been in his position, I was told a year and a half. So I gently told my host to put our report on a shelf, and not even to bother looking at it till they got the place cleaned up, because, in essence, the difference between this foundry and virtually all the reputable foundries we had seen oversees, was, in one word–cleanliness.

1985-1987 Proposal Preparation Assistance to IMI (*Israel Military Industries*)

In July of 1985, I was invited by the Washington office of IMI to help them prepare a "Producibility Engineering and Planning" proposal for an "Improved Ribbon Bridge" (IRB,) which they were bidding on, in cooperation with the well-known and respected firm, Pacific Car and Foundry (PACAR), of Renton, WA.

The IRB design was to be a basically Product Improved 1970 Ribbon Bridge, with IMI working on the multiple Interior Bay Pontoons and PACAR on the 12-meter vs. 7-meter ramps and M997 HEMET (High Mobility Emergency Medical Transport.) The design was to incorporate lessons learned by the Israelis in crossing the Suez Canal during the Yom Kippur war. Unique at that time, was, that the design team in Haifa was to be an integral part of the production process. (this concept of "concurrent engineering," would not become popular at American firms until several years later, and in France and Germany much later.)

Also, IMI was to use Finite Element Stress Analysis programs to obtain optimal thin-wall elements, optimizing the weight of each element. They proposed the use of Computer Aided Engineering and Producibility Engineering to interface at all points with hydraulic, structural, hydrodynamic and composite materials teams to assure in-process control and absolute quality assurance of the final product, including repairability and maintainability. They won the contract.

In August of 1985 I was invited by IMI to come to Israel (again) to help them prepare a proposal to the US Army Troop Support Command at Fort Belvoir, for a "New Mine Clearing System." With IMI's extensive experience in developing such systems for the Israel Defense Forces, they were pretty much of a natural. Add to the technical expertise the fact that all IMI staff members had served periodic tours of duty with the IDF as reserve officers or NCOs, they would have a pretty good, first hand interest.

IMI proposed a new system that would permit firing short, selected charges to test short stretches ahead of the vehicle, and would be re-loadable–both features not then available to our Troop Support Command. They won the contract.

Working at IMI with my spouse

In January 1986 I decided to bring along my reliable side-kick to help in a major ammunition proposal for 81 and 120mm Illumination Rounds for the US Army, to be produced at IMI's Bamifneh plant. I had arrived first to get the proposal started, and IMI managed to get me a splendid room at the Dan Hotel in Tel Aviv, right on the beach, with a view of the ocean.

By this time I had become so much of a part of IMI that they even provided a car and driver, a wonderful, former IDF soldier, who had lost a leg in the 1973 war, but who managed to drive very nicely on a car using hand controls.

My good wife, Shirley, worked not only on editing, but also prepared some of the critical Production Quality Control Plan and Drawing Change Generation Feedback & Control, that were required by MIL-Q-9858A, the Army's Quality Control Specification, as shown:

This trip really came as close to making work a pleasure as one could possibly have asked for. On the day before Shirley was due to arrive I had discovered the fantastic fragrance that emanated from the *tuberosa* flowers sold by a local vendor. So I bought a big bunch and put them into a vase in our penthouse, where their aroma infused the whole place. It was a nice welcome to a hotel, whose beachside façade had been painted, a la the artist Agam, in stripes of various colors.

And so, for the next few weeks as we worked on the proposal, we'd come down to indulge in the typical Israeli/European breakfast spread, and then prepare a sandwich to take along for our lunch, which we would eat, sitting on the stoop of the IMI building, where we worked. In the afternoons, after work, we would stroll the beach, walk the local streets and alleys, past the US Embassy that was just a block away, and occasionally stroll into town for dinner. We got to know Tel Aviv quite well, including some of the ultra-modern apartments designed by Agam.

Having completed, what was to be another successful, winning proposal to carry back to the US, I wanted to express our heartfelt thanks to the hotel staff. But when I asked the young people at the front desk how one says this in Hebrew, they were nonplussed. No one had ever wanted to extend profuse thanks–it just wasn't an Israeli thing. So they had to settle for *todah rabbah* (and a well-earned tip.) I had also learned from our breakfast waitress, who appeared frequently to refill my coffee cup, if I wanted just *od tipah?*–a few (tear) drops?

In October 1985, IMI asked me to help them respond to the US Navy Sea Systems Command IFB (Invitation for Bid) for a VLS (Vertical Launching System) Missile Canister Mark 41. They won.

In March of 1986 the US Army Missile Command at Redstone Arsenal decided to open their Improvement Program for the M72E4 Light Anti-Armor Weapon to foreign participation. IMI planned to respond with their "Anti-Armor Shoulder-Launched Weapon System" and again asked for my help in preparing another winning proposal. And so it went.

1986-My "final consulting assignments" for IMI

In August 1986 I was asked, on very short notice, to help IMI at their Ramat Hasharon facility, near Haifa, respond to US Naval Sea Systems Command RFP (request for proposal.) Luckily, I was able to engage the services of a retired Naval Captain (Norm Bull) and a retired Army Brigadier General (Phil Bolte) to help. We had just one week (from 19 to 24 August) to put together a major and, happily, winning, proposal– but at some very heavy, emotional cost! I barely made it to my plane (just as the doors were closing) and was so distressed by the IMI staff's lack of preparation and support that I wrote a long letter about what went wrong.

My close spaced, hand-written letter wound up as a three-page letter to IMI management (both in my files.) After helping IMI submit six winning proposals, it was time for new adventures.

You may well ask, after helping a company win six proposals, why I suddenly had no more invitations–not that I was sorry for one minute. Well, aside from the fact that I almost missed getting on my plane back to DC, despite the fact that I was serving as courier for an Israeli Government enterprise, I had "had it" with the arrogance and indifference of some of the staff and the way they treated two senior retired officers on my team. The two following letters to my DC contact and to the Director General, might explain the reason.

Mr. Nissim Mor, Chief, Export Office
Israeli Military Industries
3514 International Drive
Washington, DC 20008
1 September 1986
Dear Nissim :

I have just returned home from Israel an hour ago and want to send you my thoughts immediately while they are fresh reference the attached proposal in response to NAVSEA-SYSCOM RFP 54-82. I am always pleased, whenever possible, to respond to IMI's short notice invitations to help in proposal preparation.

Upon our arrival in Israel, Arnon Peled and an associate from Marketing at Ramat Hasharon were particularly helpful in speeding us through customs-in the face of the worst crowd I have ever seen at Ben Gurion airport. The arrivals actually spilled over beyond the passport clearance area out onto the tarmac. I wish someone had been able to expedite our-departure. Thanks to bureaucratic hassling, your proposal courier almost missed his plane, although we got to BGA in plenty of time.

At IMI-HD the total lack of preparation and poor support we got from people who should have been anxious for IMI to get more work, made an already difficult task an ulcerbaiting one-unnecessarily so. The lack of preparation and commitment, the arrogance and lack of cooperation by some people, was an embarrassment for me in the image it presented to my most valuable cohort, Norm Bull.

By the way, without his resourcefullness, his intelligent digging, his diligence in actually drafting whole sections of the proposal (which he really should not do as a retired Military officer) the proposal would not have been written or completed in time for submission to NavSea, and that was accomplished by the skin of our teeth.

It's all very well for us to joke among ourselves to "remember, you're in the Levant." But, if IMI wants to compete effectively and intelligently on future US RFPs in a responsive fashion, you are going to have to introduce some more effective planning procedures and get this message across from the top to the working level.

Nor would the proposal have had the technical excellence it has without the extraordinary input of the IDC staff. It was IDC, not IMI engineers, who studied the specs, prepared the work plan, PERT chart, schedules, process steps and work flow, that represent the backbone of this proposal.

The four IDC staffers were more responsive and creative than many of the IMI people. The same can be said for the secretarial help you hired to do the final typing of the corrected drafts we fed them till far too late in the game. All four of them, especially Hazel and Sarah, worked as though it was THEIR Proposal-not like the IMI staff who were largely indifferent, if not actually obstructive.

As a result of Maxi Schwarz & Chaim Berkovitch's diligence the computer generated graphics are excellent-1st class, while the copies of copies of photos depicting IMI's exceptional and exceptionally well-suited equipment, machine tools and shop facilities are 3rd class at best. They are inferior and do not do justice to IMI's capabilities. The whole reproduction process was slovenly and unskilled. Udi, who came in at the last minute from his military assignment, helped to clean up what he could.

IMI's top management, as well as its working staff has the notion that because you have in the past successfully and skillfully built complex, precision mechanisms, large and small, they can be confident that IMI can meet almost any challenge and be responsive to even stringent US DoD RFP's. That's fine–BUT, until you discipline and anticipate the proposal preparation process better, I believe that you'll lose more than you will win-despite the extensive and expensive, imported help.

You must be able to convince the proposal reader, who doesn't know about IMI, that IMI can do the job required and you must do so clearly, unequivocally and convincingly. The current proposal is fairly good and generally responsive. But, it could have been excellent and fully responsive if you had allowed yourself (and us) as much time as your world class competitors had-at least 4 of the 5 weeks between the time the RFP was issued until it was due. If your own people and IDC, as well as Norm Bull and I had been given the RFP earlier we would have been able to prepare a better proposal. A proposal going into final typing on the last day before it's due and to print the night before it's due out will always suffer from last minute oversights as this one does. To the best of my knowledge no one at IMI read the entire proposal through before it went into final print.

I recommend to you therefore a disciplined, structured preparation of IMI "boiler plate" stored in the word processors that are now becoming widespread throughout IMI. Given the time to prepare this material in a planned, orderly and unhurried fashion, you could have it available any time you needed it in the future. Just as the proposal Shirley and I prepared for Ramat Hasharon in January has reportedly already been used at least three times since as input for other proposals. That way you would be able to devote the best effort of your best people to responding to the critical technical parameters that spell the difference between a winner and an also-ran in the limited time available for responding to most RFPs.

You have some excellent capability at IMI-HD and RH from what I have seen during my three visits to IMI. As long as you are willing to go to the considerable expense of hiring outside help with special expertise and experience like (ret. BG) Phil Bolte, (ret. !JSN Capt.) Norm Bull and me, you should get your moneys worth. When we arrive your own staff should have read the RFP and be prepared to respond and, if possible, we should have had a chance to read it a few days before our departure.

Once we arrive in Israel our job should be to guide your staff in preparing a responsive technical proposal and in TAILORING the management portion to suit the current RFP. You need a photo file of all your equipment at each plant and location, as well as major process steps and finished, as well as in-process, products, with negatives and at least 5 prints of each. This file must be kept up to date at each of IMI's seven locations, by someone assigned to take the pictures, maintain a computer-stored index and cross-reference file for all locations

and issue prints on demand for proposals being prepared. Of course the quality of the photos must be such that they will Xerox well ONCE. NO copy of a copy is worth submitting in a proposal. That immediately mars the proposal.

You need current, signed resumes on file for all your senior staff, up-dated at least annually, especially now that the US military demands signed and dated 1/2 page resumes for all key project personnel,

You need clean, up-to-date pages describing IMI's organization down to the Division, Branch and Section level, including organization charts with the names of current incumbents, which can be easily reproduced for insertion at the proper place not pages that have to be cut from other proposals.

You need a Modem at each plant site to transfer data or information directly from computer to computer, not just Data-fax. Above all you need direction, so that when a proposal is to be prepared and you bring in high paid help, the respective IMI Division Chief or Technical Director calls in the staff concerned and issues a brief memo to all hands, including other divisions, to give all help necessary to the designated project engineer who is to lead the proposal.

If you do this, maybe you can turn out more winning proposals and use your consultants' time more effectively. I would be pleased to help IMI undertake such an effort—maybe in January when the call for RFPs slows down, awaiting new FY funds to enter the pipeline, please let me know how you feel about this so we can plan ahead for the task.

Sincerely, John

In hopes of getting this message to someone in Israel, who was high enough in the staff to possibly implement my suggestion, I wrote a shorter "thank you" note to the Director of the Haifa Division:

Mr. Benjamin Karmon, Director
Israel Military Industries-Haifa Division
P. O. Box 437, Haifa 31003, Israel
1 Elul/5 September 1986
Dear Mr. Karmon:

I would like to thank you for giving me the opportunity once more to work with members of your staff in the preparation of a proposal for the U. S. Government reflecting IMI's capability for responsiveness and ability. The work could not have been accomplished without the hard work and dedication of several members of your staff, IDC and your secretarial contractor.

A key role was played by Grisha Bortman, who contributed in many areas outside his nominal quality assurance responsibility. Grisha wrote whatever portions were needed, working with others on your staff in Russian and Hebrew and then creating a draft in English that answered the questions we posed. His good-natured help and bright ideas were indispensable in getting the proposal out. Indeed, I believe it is the kind of innovative ideas, which he contributed that may have helped you to get a winning proposal.

There ware many times when Norm Bull and I were ready to despair of ever getting a passing proposal out in time, let alone one that might have a chance of winning. But every time we hit bottom, Tzvi Rotem and Benny Laks came through staunchly, and never lost their good nature, until the problem had been solved and we had the answers we needed. Both Tzvi and Benny accepted more than their share of responsibility and did what they could to be accommodating and helpful in meeting all our needs, and, more importantly, in getting a proposal out in time.

The staff of IDC did an extraordinary job, working long, hard hours right along with your own staff, to create the analyses necessary to make IMI's a winning proposal. All of these, Maxi Schwartz, Chaim Berkovitch, and Jonny Vogel, did a yeoman job, and, I believe, that their computer-generated charts will go far to help you win this job.

It is, as is so often the case, not the inherent capability of IMI to make a piece of hardware to specs, within cost and time, that will win the competition, but your ability to sell IMI's capabilities to the reader who doesn't know you. IMI has ideal expertise and experience to build missile canisters for the US Navy-but without decent pictures this is hard to put across with words alone.

Finally, the proposal would not have gotten into print without the equally dedicated, hard work, and long hours put in by Hazel Packer and Sarah Seal of your contract secretarial staff, or our young friend Udi, who came in at the last minute from his military assignment and cleaned up the pagination with his wife at home before letting it go to print. Both IDC and the secretarial staff acted as though this was THEIR proposal and gave it the quality effort to help it win.

This has been the fourth proposal on which I have had the pleasure of working with IMI's staff. I hope and trust that by having this information in your computer it will make the next one easier-but not so easy that you won't need me. If I may, I would also like to urge that the next time IMI contemplates bidding on a proposal that we set up communications a little bit earlier so that we can both be better prepared to do what is always an ulcer-baiting job, with a little less grief.

By the way, without Norm Bull's resourcefulness, his intelligent digging, his diligence in actually drafting whole sections of the proposal (which he really should not do as a retired Military officer) that proposal would not have been written or completed in time for submission to NavSea. And that was accomplished by the skin of our teeth.

It's all very well for us to joke among ourselves to "remember, you're in the Levant." But, if IMI wants to compete effectively and intelligently on future US RFPs in a responsive fashion, IMI is going to have to introduce some more effective planning procedures.

I hope to have the opportunity this winter to spend a month at IMI's various plants to talk to your key people and to help set up in peace and quiet, the boilerplate you will always need for a proposal. But, by doing it when we do not have to hurry, we can set it up well, with organisation charts, resumes and a picture file, which can be up-dated every year.

Regrettably the pictures in our last proposal were third-rate copies of copies and did IMI no justice. I'd like to be able to do better next time by preparing that kind of materiel ahead of time and having it on file. I look forward to working with your staff then.

Thank you again for bringing me on board to help. With best personal regards to your wife, Anya, I remain,
Sincerely yours,
John

But, knowing that messages like this don't always make it up the ladder, I wrote to the Director General, with the understanding that it would get to him (literally and physically.)

Mr. Michael Shor, Director General
Israel Military Industries
P. 0. Box 1044, Ramat Hasharon 47100, Israel

17 October 1986
Dear Mr. Shor:
Six weeks ago I completed my fourth assignment for IMI. I am always pleased to respond to short notice invitations from Nissim Mor to aid in proposal preparation. But, regrettably instead of getting easier each time with experience gained from prior proposals, the work seems to be getting more difficult.

The lack of preparation and commitment and the poor support we got from people who should have been anxious for IMI to get more work. made an already difficult task, an ulcerbaiting one. People on one floor don't communicate with people on the next floor in the same department. We finished the proposal just in time, by the skin of our teeth, but it was not as good as it could have been. It lacked first class quality.

We had to make Xerox copies of copies of photos depicting IMI'S exceptional and exceptionally well-suited equipment, machine tools and shop facilities. They are inferior and do not do Justice to IMI's capabilities.

It's all very well for us to joke among ourselves to "remember, you're in the Levant." But, if IMI wants to compete effectively and intelligently on future US RFPs in a responsive fashion, IMI will have to introduce some more effective planning procedures and get this message passed down all the way to the working level.

IMI's top management, as well as its working staff has the notion that because you have in the past successfully and skillfully built complex, precision mechanisms, large and small, they can be confident that IMI can meet almost any challenge and be responsive to even stringent US DoD RFPs. That's fine-BUT, until you discipline and anticipate the proposal preparation process better, I believe that you'll lose more than you will win-despite the extensive and expensive imported help.

IMI must be able to convince the proposal reader, who doesn't know about IMI, that IMI can do the job required and you must do so clearly, unequivocally and convincingly The latest proposal was fairly good and generally responsive. But it could have been excellent and fully responsive

I recommend to you therefore a disciplined, structured preparation of IMI-boiler plate-stored in the word processors that are now becoming wide spread throughout IMI Given the time to prepare this material in a planned, orderly and unhurried fashion, IMI could have it available any time you needed it in the future.

I understand that parts of the proposal my wife and I prepared for Ramat Hasharon in January have already been used at least three times as input for other proposals. (Yes, I brought my wife over, at my own cost, just to expedite the work.)

That way IMI would, be able to devote the best efforts of your best people to responding to the critical technical parameters that spell the difference between a winner and an also-ran in the limited time available for responding to most RFPs.

You have some excellent capability from what I have seen during my three visits. As long as you are willing to go to the considerable expense of hiring outside help with special expertise and experience, like (ret. BG) Bill Bolte, (ret. USN Capt.) Norm Bull and me, you should get your money's worth.

Once we arrive, our job should be to guide your staff in preparing a responsive technical proposal and in TAILORING the management portion to suit the current RFP. You need a photo file of all your equipment at each plant and location, as well as major process steps and finished, and in-process, products, with negatives and at least 5 prints of each.

This file must be kept up to date at each of IMI's seven locations by someone assigned to take the pictures, maintain a computer stored index and cross-reference file for all locations and to issue prints on demand for proposals being, prepared. Of course, the quality of the photos must be such that they will Xerox well ONCE. No copy of a copy is worth submitting in a proposal. That immediately marks the proposal as second rate.

IMI need current, signed resumés on file for all your senior staff, up-dated at least annually, especially now that the US military demands signed and dated 1/2 page resumes for all key project personnel.

You need clean, up-to-date pages describing IMI's organization down to the Division, Branch and Section level, including organization charts with the names of the current incumbents, which can be easily reproduced for insertion at the proper place–not pages that have been cut from other proposals. It would also help to have a Modem at each plant site to transfer data or information directly from computer to computer–not just via Data-Fax.

I believe that IMI can turn out more winning proposals with this procedure, and I would be pleased to help IMI undertake such an effort. If it is convenient, I would like to meet with you on 16 November in your office to discuss this proposal further. I will be in Tel Aviv for

the Israeli-American Trade Week from 14 to 20 November and would be pleased to meet with you any time at your convenience.
  Very truly yours,   **John Larry Baer**

You will not be surprised, as I wasn't, that I was not invited back to do work for IMI (which I did not regret for one nanosecond, as I'd really had enough of the attitude of some of the staff, and the long, tiring flights and airport hassle). However, I found out from the DC rep. some months later, that a copy of my letter was sent to every department head for action. So I guess it must have had some impact. I wish them well. Also, I was pleased to receive the above "thank you" note from IMI. Nothing like getting an occasional "*attaboy.*"

As it turned out, as my late mother used to say–"when one door is closed, another one opens." After doing some more small jobs for IMI's DC office through 1987, I got involved with the CALS program (Computer Aided Acquisition Logistics Support), and for several years had responsibility for all the international speakers at CALS conferences here and overseas. And then there was that 3:30 AM call from France–but more of that later.

1985 "Study of the Effects of Foreign Dependency in US DoD Hardware Procurement"

So what did I do between flying back and forth to Israel? Well, one aspect of my work at IDA (Institute for Defense Analysis,) our military's dependency on foreign suppliers, was put to good use with Applied Concepts Corporation in Edinburgh, VA, and TASC, The Analytical Sciences Corporation in Arlington. The military and their civilian contractors faced a dichotomy–on the one hand we had a definite dependency on foreign suppliers, primarily Western European at that time, for parts and subassemblies of our military hardware. On the other hand, we were leery of letting so-called "critical technology" leave the military establishment, much less the U.S.

In the cited study for the Air Force Logistics Command, we found that any industrial visibility concerning the use of foreign parts was limited to first tier sub-contractors, and that the US military was really unaware of the hidden dependencies that exist in lower tiers. But, on the plus side, we found that the European end items, such as bearings, fasteners and electronic components, were available to the main contractors sooner, better and usually cheaper.

As shown in the attached (abridged) report to Applied Concepts, our travels in Europe and Asia proved invaluable in establishing contacts for potential cooperative ventures.

28 November 1985
Dear Scott:
Herewith, as promised is my report to you covering my findings at MTAG (the Manufacturing Technology Advisory Group) conference and during my visit in Switzerland to Örlikon, Charmilles, SIG, the Federation des Industrielles Luxembourgeois, and SIP with copies of some of the related letters.

In the letter to Monsieur Soisson, page 2 reflects Applied Concepts competence. The letter to Shimizu-san is strictly a fishing expedition in response to interest he expressed at the Savannah ADPA Metal Parts meeting.

Reference MTAG, my major observation was that while the Army MMT program has been virtually decimated by Secretary Ambrose, who still seems to feel that Man Tech should be paid for by private industry, Navy and Air Force ManTech programs are alive and growing, as reflected in their forward looking handouts.

In particular the Air Force expects to spend 20% of its $80 to $100M annual ManTech budget on the new RepTech program-the Repair Technology (also of interest to the Navy and Army Depots). The need here, as I expressed it to the Swiss, is for easily programmable, very flexible, computer controlled machining cells which can be used in the field by less than skilled machinists to repair the hundred and one assorted items that break and should be cheaper to repair and rebuild than to order a new one.

The Navy is also particularly interested in Coordinate Measuring Machine Accuracy Enhancement, Machine Tool Accuracy Improvement and better Tool Monitoring techniques. In-process inspection and control of dimensions and the detection and size estimation of flaws, voids or inclusions is probably the major thrust of ManTech in the near future for all 3 services.

In France, I recommended you to the *Directeur Générále* of CREATION, producers and marketers of health food products, a small, well run, growing and profitable business.

Finally, as you probably know Waterman of Quo-Tec, is looking ahead to meeting you in Farnham or at his High Wycomb office. I guess you or Jim Brock spoke to Martin Hughes of the Scottish Development Agency in Glasgow who called here for your number. I hope things work out well for your trip. Quo-Tec Ltd's logo of Technology Evaluation, Evolution & Innovation sounds as though there's a good match there to our field of interest. Good luck, my friend and have a good, and hopefully, profitable trip.

Sincerely,

John

The interesting aspect of this effort was that even engineers must sometimes think about accountability. Thus, an "Accounts Based System" developed by Applied Concepts, provided the ManTech program superior benefits tracking capability. The method was flexible in use, easy to implement, and, above all, cost effective. In addition, thanks to the, at that time, new Computer Modeling & Simulation, this method resulted in the speedy implementation of new manufacturing technologies at minimal cost and time delays.

Also, at that time, there were so few pros in the field of robotics in the U.S., that one could count them on one hand–all right, two. They were Automatix, Cincinnati Milacron, DeVilbiss, GCA Industrial Robots and Controls and Westinghouse Unimation. At that time SAIC Robotics and Automation Division was only doing "studies."

The work at TASC, The Analytical Sciences Corp., was related to our need to be able to acquire spare and repair parts faster–even those coming from OCONUS, or outside the continental United States. The problem was that it was taking 100 days to plow through the bureaucracy to even place an order–about 100 times longer than private industry.

At that time in 1986, the US Air Force Logistics Command had its engineering drawings for parts stored on 35 million, manually filed, cards! And the chances of retrieving a particular card were about 1 in 3. What TASC was trying to help to introduce to the system were aspects of Manufacturing Technology, such as Producibility Engineering & Planning (PEP,) Total Quality Management (TQM,) and Concurrent Engineering.

Companies such as Westinghouse had entire departments, with staffs up to 1200, to look at just concurrent engineering. The concept of throwing a design drawing over the transom to production engineers, and they, in turn to manufacturing, were history. Now, they would all sit around a table (even if they were thousands of miles apart and on different continents) and look at a parts design in terms of its producibility. It was an exciting time in terms of implementing new technologies–except that European and Asian firms were generally leading the march.

Consulting for Dravo Automation Systems 1987

In July 1987, Dravo Automation Systems gave me a 6-months consulting contract to help them in preparing proposals to assist the Plant Base Modernization Agency (PBMA) at Picatinny Arsenal. Their plan was to augment PBMA's staff capabilities in creating integrated information and intelligence networks, and implement the latest manufacturing technologies at the Arsenals, Depots, Ammunition Plants, GOGOs, GOCOs and COCOs. (Government Owned, Government Operated; Government Owned, Company Operated; Company Owned, Company Operated) facilities. DRAVO hoped to *"develop a rapport with key personnel directly and through Consultant* (that's me) *to obtain project information and Army long term strategic plans."*

My input to their plan was to provide intelligence and points of contact and key decision makers at Watervliet, Picatinny and Letterkenny Arsenals.

It was, for me, another "fun" assignment, in that I knew the proper contacts and the work involved primarily travelling to Pittsburgh a few times and advising Dravo about most likely approaches, and the fact that automation was not always the answer to production problems.

It was projects like this that I looked forward to when I started my consulting firm. It was a chance to utilize my expertise from 30 years of working in the munitions field and from managing the Manufacturing Technology projects for all elements of the Army's Munitions Command at Dover, NJ. Having read all of their plans and proposals and reviewed their quarterly progress reports for some 7 years, I had a pretty good idea of where they could use help to augment their in-house capabilities, and where they were quite self-sufficient, and should not be bothered.

Time out for a trip to Greece, Italy and Germany in 1987

In October 1987, at the invitation of the Deputy Chief of Mission at our Embassy in Athens, we flew to Greece for a 2-week visit. The DCM, Ed Cohen, and his wife, Ellie, just happened to be school chums from Brooklyn College, whom Shirley had run into accidentally at Washington Cathedral the year before.

Upon our arrival in Athens, Ellie whisked us off to Sunion, to see this splendid columned edifice, before taking us to their well-guarded home for lunch in their garden, overlooking all of Athens below. When we offered to take them out to dinner, she told us not to be silly, that they had a wonderful chef living at the embassy and he would prepare all our meals. The next day, I made a quick business visit to *Hellenic Arms* on *Kiffisias Avenue*, before flying off to *Mykonos*.

Brigadier *Kalkiadakis*, the Defense Attaché at the Greek Embassy here, had expressed interest in acquiring surplus medical equipment, anti-aircraft systems, tank transporters, and the "Speed Jack" used by our troops in Vietnam to turn towed 155mm Howitzers. He had also set up a meeting for us with *Mickael Tsontos*, the Director of the Helenic

Arms Industry and P. *Saccopoulos,* their Marketing Director–which resulted in some interesting opportunities for IMEC in the years to come. So as I said–it was off to the lovely island of *Mykonos.*

The tourist bureau had booked us into the charming Aphroditi Bungalows, just a short walk, through winding lanes, bordered by white-washed buildings, down to the waterfront with its splendid windmills After walking through town, stopping at an Israeli-owned jewelry shop, we headed to the fisherman's restaurant recommended to us. At the shop my wife took a fancy to a necklace and the owner suggested she wear it to dinner, and, if she liked it, she could buy it. Try that in New York! As we passed the market, there was Pete the Pelican, standing next to his owner's stall.

It was interesting to learn that you could safely put your hand in his cavernous maw, as he would just gum it. You see–that maw had to be flexible to hold his catch of fish.

That morning we had taken a local bus to the end of town, to the top of a hill, where we could walk down a goat trail to the "black beach." It was a lovely, secluded beach, and my first exposure to a "topless beach" here. It was difficult to keep my mind (and eyes) on the calm sea and gently sloping beach–but we managed. After a while a local peddler came by on a donkey with sabras, that he offered at a very nominal price, even peeling the spiny shell away with his knife, to expose the sweet fruit inside.

After enough of a lazy interlude, we clambered back up the goat trail to catch the bus, only to hear that snakes were likely to bask in sun on the rocks among the bushes, which scared the heck out of Shirley. Anyway, on the way to dinner, we saw a Sabra bush, and brave old me, I reached out and picked one, pealing away the skin with my pocket-knife. Ah the guileless stupidity of the amateur. We both got tiny thorns in our mouths and on our hands. When we told our waiter at the cave restaurant, tended by the fisherman-owner in his undershirt, grilling the fresh fish catch, he laughed and told us to dip our hands in the Greek salad he was serving, and that the oil would wash away the thorns. *Mirabile dictu,* (wonderful to relate) it worked.

A GRATEFUL REFUGEE KID'S RECOLLECTIONS

The next day we took a boat-ride to Delos. *Delos, the center of the Cyclades, has been inhabited since at least the 3rd millenium B.C. It rose in importance as the Greeks did and around 480 BC it became the center of the Delian League. The Delian League in turn kept its treasury on Delos making it a rather wealthy location. Delos became immensely wealthy and was so impressive that the Romans kept it as a free port when they came into power. It finally collapsed in 88 BC when Mithridates conquered it. What is left today is one of the most extensive ruin sites anywhere in the world. The entire island is strewn with sections of walls and marble fragments. Entire sections of the residential areas are more or less intact. The streets are still clearly outlined, many of them with sewers running underneath. Wildflowers have taken over in force in most places which only enhances the whole effect.*

Needless to say, we spent several enthralling hours wandering amongst the excavated ruins, before returning by boat to Mykonos, and walking those charming alleys to our charming, little hotel. Old ladies sat outside their white-washed homes crochetting doilies for sale.

Next morning, we flew off to Santorini. We had been booked into a B&B in the heart of town, but opted to stay in a cave on the Western side of the island, where tourists coming in by boat, rode by on donkeys, and we were treated to a gorgeous sunset, as we sat and sipped wine (without retsina) on the deck outside our room. The room had literally been carved out of the hill, but with its cool cement floor, was absolutely up-to-date in its décor, and boasted a wonderful, modern bathroom. Who could have asked for anything more?

Best of all, beside watching the donkeys carrying visitors from the cruise ships that had docked at the foot of the cliff–we saw this, aforementioned sunset.

After just one short, delightful day, we flew off to *Heraklion*, the capital of *Kriti* (Crete.) As we gathered our luggage, a young lady approached us and, hearing that we had planned to tour the Minoan *Knossos* Palace and the *Herakleion* archeological museum, asked if she could join us on our tour. Naturally, we said yes. We spent the morning admiring the frescoes that had been preserved or uncovered, whose red paint glowed as it did 2000 years ago, and I had a chance to sit on King Minos' throne. The afternoon was taken up moving from room to room at the museum, each covering a 100 year period–from the Neolithic period, covering 5000-2000 B.C., to the Palace of Knossos period (2000-1700 B.C.) and so on. Figurines, vases and tools were beautifully restored and displayed to take the breath away. It was a memorable experience.

We spent a couple of glorious days in *Aghios Niccolaus* at the *Minos Beach Hotel*, with vast lawns dedicated to sunning and a large pool for cooling off, before heading off to Rhodes. The *Acropolis of Lindos in Rhodos*, was, if anything, maybe more impressive in many ways than the *Parthenon* on the *Acropolis* (literally, the "high city") *in Athen*. But that did not in any way detract from the magnificent, impressive, reconstructed 14th century Palace of the Grand Master, overlooking *Mandraki*, the "lesser" of two harbors of Rhodes and the "New Market."

The polygonal New Market, with its huge interior court, overlooks the main harbor, where all that remains of the famed "Colossus of Rhodes," who supposedly straddled the entrance to the commercial harbor in olden days, are two columns with a bronze deer on top. All in all, the views, no matter where you turn, but especially the imposing towers and walkway to the castle entrance, are awesome and can only be described as spectacular. (Happily, the book depicting all these sites, serves as a wonderful memento and reminder of our pleasures.)

From Rhodes it was back to Athens to enjoy the sunset from *Lycabettus Hill* and *Sunion*, to walk the Agora and ancient markets, to visit *Delfi*, where once the oracle handed down words of wisdom, a bus tour of *Mycenae, Nauplia*, to marvel at the very narrow *Corinth* canal. It cuts through 63 meter (190 feet) high rocks and is 8 meters deep.

Our final stop was a visit to *Epidavros*, Southeast of *Nafplion*. Beside its ancient market square, where we strolled through the caves that served as market stalls in ancient times, it is well known as the *Sanctuary of Asclepius*, which was formed during the 3rd and 4th centuries B.C. In the center of the enclosure stood the temple dedicated to the healer god Asclepios himself. Surrounding it were colonnades

A GRATEFUL REFUGEE KID'S RECOLLECTIONS

where the sick could rest and take cures. In one such colonnade, known as the *Enkimitirion*, where we also walked, patients would spend the night seeing visions of Asclepios in their dreams and thus become cured. It was one of the early certified examples of a psychiatric clinic.

From there the bus took us to *Patras*, where we boarded a small ship for the overnight crossing to *Brindisi* in Italy, after a short stop early in the morning at *Kerkira* or Corfu. Rather than sleeping in a deckchair on deck or in the lounge, we splurged and took a cabin. Despite the splendors of Greece, when we reached Italy, I felt that we had "come home." The night train took us to Milano, in time to visit the International Machine Tool Show and meet with several of the leaders of SNIA, Pirelli and Galileo. Part of the fun of that visit was riding around town in a bus marked with the logo of their beloved soccer team, which resulted in our being mobbed by well-wishers, who thought we were part of the team.

Our next stop was LaSpezia for a meeting at Oto Melara, who were to team with Galileo and Fiat in a proposal to the U.S. Army to maintain their vehicles in Europe. Oto Melara was already busy fabricating guns and rocket launchers for our Navy, and doing a super, quality job.

The next three days encompassed a whirlwind tour of Pisa, *Firenze*, Ravenna, Bologna, Modena and Parma. (I'm just glad that I kept a detailed itinerary, so that I can recall where we stayed and when, whom we met and what facilities we visited.) On to Frankfurt, dinner at *Breuerei Glaab* in Seligenstadt, and next day a meeting at the Mainz Army Depot with their Planning Office chief, *Jürgen Schmidt*, *Heinz Theuerkauf*, Contract Administration Chief (hard to have a name that means "buying expensive" in that position) and the Commanding Officer, Colonel Bill Littlefield. (There's no such thing as a simple pleasure trip when you're running a consulting firm–if you can make business contacts along the way.)

Another quick detour via Baden-Baden to visit a cousin in *Bad Herrenalb*, and on via the sleeper to Paris for a meeting with *Madame Catherine DuPont-Gatelmand* and the Commercial Manager for Europe at *Renault Machines Utils*. It makes me tired just to recall that one-month jaunt across Europe, returning home just in time to vote on November 4th 1987.

DGA International & Oto Melara 1988 to 1989

Sometime around January 1988 I had been approached by Craig Musick, the General Manager of DGA International, to assist them in working with Oto Melara in La Spezia, Italy & FIAT in preparing a proposal to "Maintain American Combat Vehicles in Europe." The two-year consulting agreement would also encompass other proposals to support the military's "Heavy Force Modernization."

Working with Oto Melara was probably the most professionally fulfilling experience of my career. The firm was run by engineers, from *Presidente Ingeniere Sergio Ricci,* down to the shop managers. Whenever the engineers putting the proposal together, with help from a FIAT and a Galileo representative, huddled to discuss *en Italiano,* how to put their best foot forward, I would just ask: *"per piacere; parlare lentamente,"* (please, speak slowly) and they would slow down enough so that I could get the gist of their concerns.

As a result, they put together a fantastic proposal, certain to meet the needs of the U.S. Army. There was just one problem–FIAT. With their far less tidy manufacturing facility, they insisted that the price arrived at by Oto Melara be almost twice as high. So, despite FIAT's very fine test facility, and Galileo's exceptional electronics rehab layout and staff–they just priced themselves out of contention. It was a sad point in our work, because with their reputation and their other work, building guns and rocket launchers for the U.S. Navy, they would have been a "shoo-in." So our best intentions were to no avail. Nevertheless, Dr. Ricci, hoping against hope, that their reputation would overcome the high price, took us all out to lunch at *"Parigi,"* his favorite restaurant in the countryside near La Spezia. The *prima collazione,* or first course of that unforgettable meal, was a *Risotto,* which Dr. Ricci asked to be topped with *fungi nero*–or black truffle. So the waiter brought out one of these rare truffles, about 2" in diameter, and shaved some over each of our steaming plates. The aroma was something I will never forget.

Foster Miller, Waltham, MA-1988

In March, 1988, based upon some of my experience at LWL and what I had learned from my visits to Italian robot manufacturers, I was invited by Foster-Miller to come to Waltham, Massachusetts, to help them with their TALON robots and LAST Armor projects to better protect our armed forces. They were also concerned with fine-tuning the technology for human factors and human performance so it is safe, easy and efficient to use.

Westinghouse Waste Technology Services, Waltz Mill, PA
In view of the ever-present problem (even to this day) of destroying toxic waste materials with a minimum of secondary waste products, I had been working for some time with the Westinghouse Waste Technology Services Division to try to market their highly effective "Plasma Arc Pyrolyisis" system. And yet, despite the support of some of their top management, it somehow never did get off the ground, which is unfortunate. My letter to Aris Melissaratos, their Vice President for Quality, and to the Director of Remediation Services, shown on the next pages pretty much tells the sad story.

1989, a busy year in America and France

In June of 1989 Dynatech Communications of Woodbridge, VA, asked me to assist them in enhancing their production operations–everything from product design, to fabrication, testing, packing and packaging. Dynatech had been under some pressure to "perform, or be divested." It was an interesting short-term, 16-week challenge, but the best was yet to come.

On Friday morning, July 21st began a momentous portion of my consulting career. I had a call from the French Embassy to come over as soon as possible to discuss an urgent need by the French company *Navfco Logistique,* for some consulting support.

Well, after a quick trip to their offices, I ran home to prepare a FAX covering some aspects of my background that they would need to know about, and sent it off to France. (Remember, this was in the days before e-mail.)

Monday morning, at 3:30 AM our phone rang. "*Allo, Monsieur Baer.* This is *Claude Legat in Paris.* Can you come right away, *toute de suite?* We need your 'elp right away." I understood that it was already 9:30 in the morning *á Paris,* and foregave M. Legat for waking us. So I told him that we would catch a flight to Paris that afternoon and for him to meet us at *Charles de Gaulle* airport the next morning. (Again, that was in the days when you could catch a quick flight, and flying was a pleasure–not a hardship.)

As planned, he met us in Paris and quickly transferred us to Toulon, for the start of what would be many years of happy work, helping *Societe Naval Francaise de Formation et de Conseil (NAVFCO.)*

I guess that our $ 100 an hour fee seemed modest to them, because they never balked at paying for the numerous hours we worked on their behalf. And, indeed, Claude Legat, their chief, and I, remained friends for many years.

The work with Navfco brought me in touch with *ilce* Iniatives, who, following in the footsteps of the CALS Europe '93 International Conference in Berlin, led to *ilce '94, '95 and '96,* to expose French and other European executives to the benefits of "Integrated Logistics and Concurrent Engineering." In fact, one of my first assignments was to represent *iXi/ilce* at CALS Expo '94 in Long Beach, CA, and making final speaker preparations for *ilce 95,* which was to be held at *Espace Saint-Jacques* in Paris from 1/30 to 2/3/1995 in Paris.

With the contacts I had made in the CALS program, with American industry leaders, and those from Europe and Asia, whose speaking assignments I coordinated, I proved to be the perfect foil to bring some of these speakers to Montpellier and Paris. It wasn't always easy to get the right people to provide a well-balanced agenda, and so, on the Sunday night before *ilce 94,* the *Président, Michel Ladame,* told me over dinner: "I don't want to see you tomorrow."

To say that I was non-plussed and taken aback would be an understatement. When I asked him, why? He said that I had been working so hard, that, after breakfast he wanted me to go next door to pick up the car he had reserved for us and drive through the *Camargue Delta* region of *Provence,* to Arles to see the Romanesque ruins that date back to the time of Julius Ceasar, to chat with *les Arlésiens* as we walk the town after *dejeuner* (lunch), Aix-en-Provence, and Les Baux-de-Provence. Other splendid sites in the area, such as *Avignon* (famous for the P*ont d'Avignon),* Nimes and *Marseilles,* would wait for later visits.

Well, we did as directed, driving the lovely, winding roads down to the harbor at Arles, walking the quaint streets, after a wonderful repast at a little *patisserie,* and then headed up into the hills for a walk through *Les Baux,* that we really hated to leave, it was so charming. But we had to return to give our report over dinner that night. (When I say "we" in all these tales, it's not the "royal 'we,'" but reflects that my good wife, Shirley, was at my side, helping me through problems, wherever we went.)

To appreciate the international flavor of these *ilce* conferences I have to note that our distinguished speakers were assembled from Austria, England, Finland, France, Germany, Holland, Italy, Japan, Portugal, and, of course, the U.S. And while the accompanying Expo was not as elaborate as we expect in the U.S. it provided a showcase for all the European leaders in the field, including *Aérospatiale, Alcatel-Israel, Cegelec, Giat, IBM, OBS GmbH & Siemens Nixdorf.*

Fortunately I had some help in making all these speaker arrangements, from my partners at IMEC, Dave Bettwy, formerly at the Bureau of Standards, & Naomi McAfee, who had retired as Director of Concurrent Engineering at Westinghouse.

Thus armed with contacts, we were able to bring in topnotch speakers for *ilce '95* at the Expo Center in Paris, and *ilce '96* at the Hôtel Mercure in Paris in October 1996. For *ilce '96* we were able to add speakers from Belgium, Brazil, Spain and Sweden.

The featured speaker for this event was Jan Hines from Lucent/Bell Labs from Orlando, Florida, who provided an eye-opening discourse on "How the Japanese Industry Operates." Remember that at that time the Japanese were leaders in Human Resource Management, Quality Systems, such as *Kaizen* or Continuous Improvement, Quality Circles and Statistical Process Control, as well as Computer Development and Use, both to improve Total Quality in Manufacturing and to improve efficiency in the Office.

At most of these conferences the audience was attentive and appreciated the quality of the speakers we had provided. There was, however, one occasion that proved to be a challenge. *IXi* had asked me to set up a one-day seminar on Concurrent Engineering. Our associate, Naomi McAfee, having directed this work at Westinghouse, was the ideal candidate; so we brought her to Paris, and I proceeded to introduce her to the audience, which was composed largely, of men.

Typical chauvinists, their first reaction was aghast that they should be lectured to by "a woman?" I had to quickly explain that Mrs. McAfee had led some 1200 logistics engineers at Westinghouse. So they finally settled down and listened as Naomi explained that the old ways of tossing design drawings over the transom to the manufacturing engineer, who, in turn, threw them over the transom to the production folks–was a thing of the past.

The problem with that method was that a fine design might be hard to produce, and so, it made better sense for the designers to present their concept to the engineers, fabrication folks, inspectors, and the customer, around a table, where they could provide input, so as to assure the design's producibility.

Well, the morning went well, but after lunch when Naomi said that we were going to start by playing Scrabble, I thought we would have a revolt. "Is this what we paid for? To play games?"

So, again I asked for patience, as Naomi asked who was the best Scrabble player in the room, and got five volunteers. She split the remaining twenty people into five groups of four, directing that they should each play competitively. Of the 5 "experts" she asked who thought that he was really good, and assigned him to play by himself, using any tiles he chose. The other team of four was directed to work together, letting each player play to his strength in terms of vocabulary, visual sense, etc. Then she gave them 15 minutes to play.

At the end of that time, she got scores from each group. The normal teams got scores between 90 and 120. The expert got 145; but the team who could work together, got 210! Aha! The gist of Naomi McAfee's presentation was as follows:

CONCURRENT ENGINEERING
(As an Element of Total System Life Cycle)
Concurrent or Simultaneous Engineering is a *Management Tool*
For Concurrent Product and Process Development.

The Purpose for using Concurrent Engineering is to do away with time-consuming & error-prone Sequential Development Process. Instead, we wish to INTEGRATE the Design of a Product with all the related Processes required to get it to the Customer.

Review of the Production Processes must include Product Design, Tool Design and Gaging, Engineering, Manufacturing, Purchasing, Product Marketing, Post Delivery Maintenance and All Related Commercial Support Activities, including *Customer Representatives!*

Critical Questions:
What does the Customer Want?   The airlines? The airports?
What does the Customer Really Need?   **Have we asked them?**
What is the Competition Providing?   What works in Europe? in Asia? in the U.S.?
How does your Current Product compare to the Competition? You'd better know that!
Can you provide a better product at the same price? or
Can you make the product so much better the customer will pay a higher price for the product?
What are the Technical Hurdles to meeting Customer Needs?
What are the Potential Trade-Offs?
How to Start the Concurrent Engineering Process in Factory or Office!

Conduct top-down analysis of the Product Design Process using a team of knowledgeable engineering and production specialists, sales and marketing personnel and customers.

Companies all over the world, (like Airbus, Boeing, McDonnel-Douglas and Northrop) including our audience, quickly learned that it was possible to design and build a flying model of a brand new aircraft without first building a model or mock-up. In fact, during my last trip to AVSCOM, (the US Army Aviation Systems Command) in St. Louis, they showed me, with pride, a new fighter that had been built and assembled with such accuracy, that everything aligned perfectly and needed no adjustments.

So why have I bothered you by making you look at these 20+ year-old letters/FAXs? For me it's a nostalgia thing, knowing that, maybe, my experience has done somebody some good. The EsoCE Newsletter FAX gives you some idea of the global impact of this new concept, and that in those bygone days, the FAX machine, rather than e-mail, was the *modus operandi*. This reminds also that, when we were setting up our trip to Russia, I would call the Institute's phone in Kiev, and when they answered, say: *"FAX pajaluista,"* (FAX please) and they would switch from phone to FAX, so that I could then transmit my FAX.

To show how well we succeeded in helping start the *European Society of Concurrent Engineering*, read the letters shown in the middle of this book:

Going to jail for *Black Diamond Enterprises* in 1990

In February 1990, John Robinson, President of *Black Diamond Enterprise,* in nearby Forrestville, MD, just across the Potomac from us, invited me to help improve the productivity of his small sheet metal business, that was fabricating food racks for McDonalds and other dispensers of fast food. Fortuitously he had opted to attend a meeting of the local chapter of the Institute of Industrial Engineers, when I was giving a talk on improving productivity through better plant layout. As it happened, he was planning to move his shop to another nearby location and buy some new equipment.

When I visited his shop, the machines were pretty much scattered throughout the place, in the order they were received–not in the order of the process from cutting to welding to heat treatment. The first thing I did was to identify each piece of equipment, including its capacity, age, capability and cost, and the same for the qualifications of the staff running the machines.

This was a great opportunity to utilize what I had learned over the years and to apply it to a small, minority-owned business. What I proposed was as follows:

To conduct an analysis of their facility and operations and propose ways and means to improve their productivity, potential for growth and entry into the military market.

As a first step I proposed to realign the layout of the equipment, providing an orderly flow for material processing and making provision for orderly storage of raw materials and work in process–in one door, out the other.

Second I proposed a cross-training program for BDE work crews so that they could not only adjust and maintain their own machine, but also any other machine in the shop.

As a 3rd step I proposed that the increased productivity, which could be expected from smoother operations, be translated into productivity bonuses for the staff. I proposed also to provide an incentive to avoid scrap and rework (which again translates into improved profit and potential for bonus.)

To do this I believed that he needed to first reach out to other companies that have demands similar to his current clients; then to hotels, hospitals and other organizations that use similar equipment to that which he makes currently for McDonald and Pepsi.

The next step is to register BDE capabilites and interest with the Small Business Administration, Commerce and Agriculture Depts., other Government cafeterias that use racks, such as the Pentagon, Andrews Air Force Base; then the Depot Systems Command and, of course, the Defense Reutilization & Marketing Service to identify the kinds and type of equipment BDE might want to buy to expand their output capability.

The entire process must be planned carefully and not hurried. If there is an interest in having BDE grow, it MUST be done slowly and incrementally, allowing the company to digest each new capability before moving ahead. I suggested also that he limit growth to remain a small business. The advantages, in my opinion, vastly exceed the disadvantages of bigness.

In summary, I proposed that BDE try to expand their current small niche of a specialty market to providing a quality product to additional customers and, in the process, expanding their capability and capacity to produce other similar, related and later, even dissimilar products. The secret to their success would lie in their ability to maintain overview and control of the operation as it grows.

Before setting up any process improvement steps and cross-training the staff, however, I opted "go to jail" for him, by visiting the Ohio State Penitentiary in Youngstown, where the inmates were under contract to fabricate similar racks. For a sheltered government worker, that was quite an eye-opener. Sure I had read books that involved lawyers visiting inmates in jail, but doing it myself was different. It turned out that the inmates were really quite productive, careful in their work, and surprisingly enthusiastic. I was allowed to take pictures of the equipment at work breaks, but not if there was an inmate nearby. We were able to apply some of the best practices I learned there, and to arrange for an orderly layout of BDE equipment in the new facility. It was, all in all, a very satisfying assignment.

1991–The Soviet Institute of Metrology and Standardization

In December 1990, *Dr. Vladimir Novikov,* Deputy Director of Sciences at the *Soviet Institute of Metrology and Standardization* wrote to the then president of our local chapter of the Institute of Industrial Engineers, Bob Charlton, asking for help. (*see letter below.*) I thought that this might be an interesting opportunity to see Russia, not that I expected to make any money–in fact it cost us. But anyway, arrangements were quickly made. I went to the Soviet consulate to get visas for us and in May 1991 we flew to *Moskva Sheremetyevo*

A GRATEFUL REFUGEE KID'S RECOLLECTIONS

International Airport, where we had to transfer to *Vnukovo*, the regional airport for our flight to *Kiev*.

Needless to say, the signs were all in the *Cyrillik* alphabet, and while *Sheremetyevo* was not too bad, on the order of one of our small town airports, *Vnukovo* was the pits. It was small, dirty, poorly lit–in one word: primitive. At one point I thought I wanted to go to the toilet, but I took one look and ran out. The filth and smell were absolutely unbelievable. In fact, I took a picture–and then threw it out, it was so awful. Fortunately out connecting flight did arrive in short order, albeit two hours late, and we joined the stampede of baggage-laden *Russkis* to get on the plane and fight for two seats.

Once we arrived at *Borispol* airport in Kiev, we were met by our host, Dr. *Vladimir (Volodya) Novikov*, bearing the traditional bouquet of flowers to welcome us. He brought us to the *Kozatsky* hotel, located right on the main square, where we were greeted by machine-gun toting guards, who initially blocked our entrance to the hotel, till *Volodya* vouchsaved us. Well, it wasn't all that bad–the elevator worked and the room was a passable two-star hotel room.

After dropping our luggage, we were taken to the Institute, where we were greeted by a room full of distinguished listeners, such as the Directors of the Minsk Watch Factory, the Kiev Chamber of Commerce, the Kiev Polytechnic Institute, the head of their CAD Department, and the U.S. Consul General, among others. Happily, all these good folks understood English quite well, and asked good questions. After the less than appetizing arrival, this was a welcome time and seemed to be beneficial and fruitful for all concerned.

Naturally, *Volodya* brought us to his apartment on *Svyatoshinskaya Ulitsa* for dinner, with his wife, *Natasha*. In typical Russian fashion, they rolled out the red carpet for us, with caviar and blini and Vodka and frequent trips to the little ice-box on the balcony of their small apartment to fill the table with goodies.

Years later we were to experience similar hospitality when our son, serving at the embassy in *Skopje, Macedonia* took us to visit a Serbian farmer.

Next morning we got a tour of the Institute's brand new building, where *Volodya* proudly showed us their "new" quality control measuring equipment–which resembled stuff that we had thrown out twenty years earlier. But who's to argue with folks whose AK-47 survived the Vietnamese mud better than our M16's and who launched *Sputnik* years before our Apollo.

Thus, we had the afternoon free to visit the very moving memorial sculptures and plaques to commemorate the trials, tribulations and suffering of Soviet soldiers and civilians in World War II, on the high bank of *Dnieper* River. The sculptures were so moving that I take the liberty of enclosing some of them in the middle of this book.

Of course, we also visited *Babi Yar,* the site of the massacre of thousands of Jews, where a musician now sits and plays typical Russian laments on his Bass *Balalaika,* which I mistook for a Domra. (I don't know if it was the Russian 3-stringed or Ukrainian 4-string version, but it is akin to a lute, and is used to play mournful tunes.)

In the evening we went up to the hotel's roof terrace, where a nice band played American style dance music and a friendly waiter supplied us with coffee and cakes. When he finally was ready to go home he bade us *spakonye noche* (good night), and when we asked him what we owed, he said "von dolla." Those were the days when a dollar was worth a lot in Russia.

I should note, in passing, that I think our visit helped the Institute in that in April 1993 the Russian Parliament adopted the Law on "Assurance of Measure-ment Uniformity". The Institute made good progress in the work of putting the Law's principles into metrological practice. Today, it is accredited not only as a scientific organization, but also as the State center for testing, certification and verification of measuring instruments, as well as a scientific and methodological center for the Russian calibration system, etc.–somewhat akin to our Bureau of Standards, now the National Institute for Standards and Technology in Gaithersburg, MD.

Since we had planned for a couple of free days before returning to Moscow, we flew to St. Petersburg, the Russian Venice, and visited the world-famous *Hermitage* Museum. We had arranged, through Russian friends back home to rent an apartment for a few days. The owners of the apartment were happy to vacate it and move in with friends, as it meant earning some hard to get Dollars. As it turned out, you had to turn on a small, overhead water heater in the kitchen to get hot water to wash, and the bathtub was so grimy and grungy that neither of us was willing to sit in it. Ah well, at least the beds were clean and living room comfortable.

We opted to eat only at so-called "dollar restaurants" situated along the several canals that cut through St. Petersburg–a) because they were clean and safe, and b) they all had nice, clean, imported, German toilets. You quickly learn what's important. We also learned how to shop at the local 5-story book store, where you had to find the book you wanted, then stand on line to pay for it, and then fight the crowds to actually retrieve your purchase.

We encountered an interesting episode also, upon our return from visiting Catherine's island getaway, her summer palace, called *Ekaterinhoff*. When we got off the boat, rather than trying to fight our way back on the very well-run, but very crowded subway, whose trains ran punctually every 4 minutes, as noted by a countdown clock (in 1991!–what took us so long to catch up?) we looked for a taxi. There was an empty one sitting at the curb, but when I approached it, the driver, noting my American clothes, said in surly fashion: "I am bizzy." So I pulled out one of those magic dollar bills. When he saw it, he said: "vere you vant go?"

So we got to the Hermitage and went up to a window to buy tickets. As soon as we got up to the window, they slammed it shut. So we went to a second window–same results. At the third window I was prepared– as the surly ticket seller was about to slam the window down I stuck my guidebook in the way and refused to remove it till she sold us two tickets. So we saw the exhibits–but after having seen the Louvre, and the magnificent art museums in London, New York, Frankfurt, the Vatican, etc–we were disappointed. The lighting was poor, the maintenance awful, and we left after a quick walk-through to find the Great Gate.

Kiev Great Gate (*from Wikipedia*)

The **Golden Gate of Kiev** (Ukrainian: Зо оТі ВороТа, *Zoloti vorota*, literally 'golden gate') is a historic gateway in the ancient city walls of Kiev, the capital of Ukraine. This gateway was one of three constructed by Yaroslav the Wise, Prince of Kiev, in the mid-eleventh century. It was reputedly modelled on the Golden Gate of Constantinople, from which it took its name. In 1240 it was partially destroyed by Batu Khan's Golden Horde. It remained as a gate to the city (often used for ceremonies) through the eighteenth century, although it gradually fell into ruins.

In 1832 the ruins were excavated and an initial survey for their conservation was undertaken. Further works in the 1970s added an adjacent pavilion, housing a museum of the gate. In the museum one can learn about the history of construction of the Golden gate as well as ancient Kiev.

In 1982, the gate was completely reconstructed for the 1500th anniversary of Kiev, although there is no solid evidence as to what the original gates looked like. Some art historians called for this reconstruction to be demolished and for the ruins of the original gate to be exposed to public view.

The "great" or Golden Gate of Kiev (*reconstructed*)

NOTE: The extant reconstruction of the gates has been dismissed by art historians as a highly controversial revivalist fantasy. A picture I took of the Great Gate, as a memento, is shown in the middle of the book.)

And so it was on to Moscow, where we made a Bee Line to the stand selling bottles of Irish spring water and each gulped down a full liter, hardly stopping for a breath. We were met by our host, Dr. *Serge Youmashev*, who drove us to the hotel where we had reserved a room.

It was a lovely, new hotel, with just a few cars in the parking lot, but when we got to the desk, despite my letter confirming our room, the woman boldly told us that they were full. I had thought about slipping her a few dollars or some nylons, but my escort was too cowed by the typical bureaucracy, that we left and he took us to their dorm, where we had to climb five flights of stairs with our luggage to a small room.

Embarrassed as he was, he offered to take us to his home for dinner, where his wife, *Lyuba*, tried to offer us something on their obviously limited salary.

Not having been able to turn in our passport at the hotel, for the police to check it, I knew that we were in trouble. We went to the police station, but the only way we would be allowed to leave the country, was for the Professor to leave his papers as *bona fides*.

We briefed *Dr. Nicolay Ivanchuk*, the Rector of the All Union Institute of Standardization & Metrology and his deputy, *Dr. Olga Rakhovskaya*, who then escorted us on a tour of the Kremlin, Red Square, the GUM department store, and other Moscow sights. By then it was time for dinner and we went to the *Rossia* Hotel on Red Square, where as the "rich American" I treated the whole crew to dinner of caviar, *blini*, steaks, champagne–the works. The dining room was only partially filled, and so, since the band was playing nice "American" dance music, we got up to dance. When the band stopped, I went up to ask where they were from in the states and they admitted that they were Russian and had just learned to copy our music–very well, I might say.

When I asked to pay the bill, the dinner for seven of us came to just $37, and when I gave each of the staff a dollar for having served us so well, they actually kissed my hand. Wow! Who would have imagined it? That night I couldn't sleep and was sweating so badly that the sheets were soaked. I had passed blood that day and knew something was wrong. So here we were on the 5th floor, with no idea how to call 911 or even the necessary *dvushka, the 2 kopek* coin to get an operator, who probably spoke no English. Shirley, scared out of her wits, just pleaded with me to hold on till we got to Frankfurt next day, where they could take care of me.

Well, I made it through the night, got to the airport and, thanks to our authenticated Visa, through customs, so that our professor could get his papers back. I think he was relieved to see us go. Just ahead of us was another couple from the States, who were missionaries. When they got to the Customs officer he asked where the computer was that they brought into the country? They told him that they had left it at the mission they had visited. So they were told that they had to pay a $ 400 customs fee! They pleaded that they had left almost all their cash with the mission–to no avail. Fortunately there was an ATM at the airport, open even on Sunday, so they could get the cash. It was an experience that has stuck in our memories all these 18 years.

It's also worth noting, *en passant,* that during our dinner at the *Rossyia*, Olga asked Shirley what she does for work? Although we had not mentioned to anyone that we were Jewish, having been taken aback by the virulent, anti-semitic displays at a Leningrad Metro station, Shirley admitted that she worked for a Hebrew School. Olga, in total surprise said: "Oh my, I'm Jewish too."

It had been a costly trip, and quite an experience, but, boy, were we glad to be back on a Lufthansa plane that got us to Frankfurt, and in short order back to the good old, US of A. Just for giggles I'm going to include our "observations." I don't ever have to go back there.

OBSERVATIONS FROM THE SOVIET UNION

Two days ago sitting at Moscow police headquarters, facing an intransigent, obdurate KGB apparatchik, posing as the Visa officer, we could not have imagined sitting out here on our deck at home, safe and sound, listening to the birds and allowing our lungs to purge themselves of 9 days accumulation of automobile and truck exhaust fumes and omni-present cigarette smoke.

What was our crime, you say? Someone had slipped up getting us a hotel reservation and the hotel in question refused us a room. So our host graciously offered to let us stay at the dorms connected with the Institute we were visiting (after vainly trying to argue with another intransigent hotel apparatchik.) The room was clean, with a decent bath and kitchen-albeit on the 5th floor of a walk-up (a bit much for two Senior Citizen types.)

When you visit the USSR, as in much of the world, you turn in your passport to the hotel when you register; they log it in with the police, and that's all. In the USSR they also certify the days of your visit, and if you DON'T stay at a hotel you must go to the police, fill out a long form and plead-before they'll stamp your visa. In this case the agent and everyone of his henchmen/women down the line claimed our visa had been granted by the Institute's headquarters and THEY had to request our permit. [Somewhat like visiting NIST/Bureau of Standards, and then insisting the Dept. of Commerce is our host.] Another vignette-we met a Christian missionary who had left his computer with a Russian church; he had to pay a $400 tax to be allowed to leave! Well, we're home, much the worse for aggravation, polluted lungs, dehydrated bodies (you DARE NOT drink the water anywhere & you can't always get bottled water.)

Aside from the long lines wherever anything of value was being sold & the humungous crowds in the Metro in Kiev & Leningrad, the vast number of cars with broken windshields being driven around, the miserable roads and the filthy conditions we observed in certain places, what were some of the GOOD observations?

Everyday people and our hosts were invariably and extraordinarily warm and hospitable. Our host's wife in Kiev invited us to their home for dinner; we got to see an opera, as well as numerous monuments, Babi Yar, onion-domed churches and other sights. Although food is rationed (30 eggs, 1 Kilo flour, 1 Kilo butter/month) the table at dinner groaned with food, vodka, champagne, caviar. In Moscow our host's wife not only prepared a fine feast for us on an hour's notice, but even packed breakfast for us, knowing that there were no dining facilities in or near the dorm. You couldn't have asked for more.

The Hermitage in Leningrad may have had more paintings than the Louvre and some gorgeous royal chambers, but personally, we'll probably stick with the Louvre and the Musee d'Orsay. The one sight which can't be seen anywhere else in the world, is St. Basil's onion domed cathedral on Red Square.

Magnificent is an understatement. The Metro trains run every 4 minutes (2 minutes during rush hours) and a digital clock at the platform tells you how long ago the last train left. Watch out for the doors! They close with a snap-no spring-back. But one has to put it in perspective. We'll probably be able to help the Institute of Metrology

and Standardization who invited us to set up a Total Quality Management training program. Would we ever go back? NOT VERY LIKELY!
by Shirley & John Baer 6/9/91

P.S. there was after all, a happy outcome to our visit. In November Dr. *Serge Yumashev,* Director of the Scientific Institute notified us that, as a direct result of our intervention that Booze, Allen, Hamilton and Excel developed partnerships with the Institute. As a result of our lectures they also realized that they need to set up training centers for quality control in the *Ukraine, Byelorussia, Moldova, the Caucasus,* etc, to provide certification that is now recognized everywhere.

Swiss Machine Tool Builders Assn., Zurich

Following a meeting with the Swiss Machine Tool Builders Association in Zürich to discuss their implementation of CALS, resulting from their involvement in previous CALS Expos, we planned to follow up on a previous invitation from Swiss Telecom, who were also interested in implementing CALS. As a result of their inquiry, I sent them the following letter:

Herrn Markus Gertsch, gema2@gd.swissptt.ch
Swiss Telecom, GK-EP, Tel. +41-31-338 05 27
Viktoriastr. 21 Fax: +41-31-338 31 07
CH-3030 Berne/Switzerland
19 September 1996
Sehr geehrter Herr Gertsch:
Thank you for your FAX and e-mail of 18 September. As I noted in my reply, we would be happy to meet with you on Tuesday morning, October 8th, at 9 AM. I would presume that our
  meeting would last 2 to 4 hours, depending upon the amount of questions and discussions generated.
We can certainly be available until you have to leave at 4 pm.

My wife/associate and I would be pleased to meet with you as well as anyone you care
to invite from the Unisource parent companies or any Swiss Telecom subsidiaries. In fact, their
participation would be helpful since *concurrent product and process development is predicated*
upon bringing your customers and top management into the process from the start. Since you know
who is involved with Swiss Telecom, both up and down the line, it would be far better for you to invite
them than for us.
My plan is to discuss with you and your associates:
the need for electronic commerce and other elements of CALS, in order to be viable in a
virtual enterprise environment;
the problem areas & pitfalls in adopting new technologies like Electronic Commerce (EC);
the potential benefits to be derived from implementing EC & CALS and promoting
concurrent product and process development;
the importance & impact of standards, incl. product data standards
the concept of Swiss Telecom as a linchpin or hub for Swiss EC;
the potential for Swiss Telecom as a global teaming partner; and
finally the challenge and dangers of Y2K-the impact of 2000 on computers!
I will bring my laptop computer with me, which will have my tutorial slides on it from Seoul
and Athens, plus some I will generate especially for this meeting. Can you have an LCD projector available for me to plug into? Would you also be so kind and reserve a room for us at a nearby hotel and let us know the name and location. We will be driving up from Portofino.
We look forward to meeting with you.
Very truly yours,
John Larry Baer
JOHN LARRY BAER, P.E, CALS Expo '96 Plenary Chair

As it turned out, this was an extremely productive meeting, not in the sense that it resulted in a contract for IMEC, but in the good will created and in Swiss Telecom's subsequent involvement in the CALS program. (and it provided a very nice visit to the old town of Bern for us.)

Explosives Detection Systems study for our airports (1996)

In the fall of 1996, I had the chance to participate in the fascinating work of serving on the *National Materials Advisory Board Panel* on the "Technical Regulations of Explosives Detection Systems." What this meant was studying our vulnerability to smuggling explosives onto airplanes and finding technology to thwart such attempts. We looked at X-Ray, Nuclear, ElectroMagnetic and Trace technologies for passengers and checked & carry-on baggage.

There were already several commercial systems on the market, but most, except for a German unit, had a sizable footprint, and would require reinforcing the floors at many of our smaller airports. Chief among the contenders was the InVision CTX-5000, but, like many of the contenders, tended to a high false alarm rate–something with which airport security personnel could not cope.

At that time, San Francisco Airport personnel had to use human intervention to resolve a false alarm rate of as high as 33%. (The unique problem at SFO was thought to be the prevalence of large 20 x 20 x 20 metal, duty-free boxes being shipped to the Philippines, that could be packed with virtually anything.)

We also looked at the concept of *nuclear quadropole resonance*, which is effective in detecting Nitrogen, making it useful in detecting sheet explosives such as *RDX and HMX*. Of interest to the panel was also the fact that, while InVision had spent two years developing the CTX-5000, they had very little production experience. Since that time, they have, of course, given a lot more attention and staffing to configuration management, quality control and control checks to assure reliable systems performance. We've come a long way since then.

In earlier years, back in 1981, I even had an invitation to follow the Undersecretary of the Army, James Ambrose, with a brief banquet speech to *Ballistic Missile Defense Program Managers*

Rio Library InfoTech Conference, *Rio de Janeiro, Brasil, 13-14 November 1997*

I had never been to Brazil and hence was very pleased to be invited to speak to the International Library InfoTech Conference to be held in Rio de Janeiro. The company making the arrangements offered to pay my fare, hotel and provide a modest stipend ($ 3000, as I recall.) After considerable research on the application of Information Technology to the Library System around the world, I created a 13 page paper and 32 Power Point slides to present to the conference. (see Appendix)

Though somewhat afield from my area of expertise, the Mellon Foundation had asked me to give a talk on "An Engineer's View of the Information Society and its Relevance to the Library System" in Rio de Janeiro. Since I was hardly a library maven, I did quite a bit of research for my paper and made sure to have it peer reviewed by two librarians and two industrial engineers to make sure that my paper would be closely library related. I used these as my two opening slides:

One of the interesting aspects for me, was when I checked in to the Rio Palace Hotel, conveniently located across the street from *Copacabana* Beach, the clerk gave me my room key and a safe key. Then she gave me a caution to the effect that she wouldn't let me go out on the street until I locked up my valuables in the room safe and showed her the key!

Well, I did as she suggested and went for a walk, checking out *Copacabana* Beach, and then walking across the isthmus to check out the famous *Ipanema* Beach. Along the way I noted that apartment houses had heavy steel bars from the ground to the first floor to protect against thieves, with a guard patrolling both inside and outside the bars. That was a bit of an eye opener.

On the other hand, I had one very pleasant experience, after giving my talk and paying a taxi $ 12 to go to the mountain to view the Cristos Monument that is a well-known landmark. Coming back down from the cable-car ride, I asked one of the guards if there wasn't a local bus that would take me back to *Copacabana* Beach? He assured me there was, and so I took the # 456 bus when it came along. Not knowing the exact fare, and since the Spanish I knew was not Portuguese spoken in Brazil, I gave the ticket taker 2 Cruzeros, after telling him where I wanted to go.

After a few minutes he came and tapped me on the shoulder, asking: *Senhor, deusch?"* (i.e. you gave me two?) When I answered, *Si*, he gave me one back, and refused my offer to keep the other. Poor, maybe, but very honorable. The bus was filled with working class folks, one young man taking off his shirt, as it was warm. After some time, several of the people who had heard my destination, gave me to know that mine was the next stop. That gave me a good feeling, and I was glad that I had taken a bus, rather than a much more expensive taxi. As for the "Girls of Ipanema"–from what I saw, I thought they were highly over-rated. I'll take the Riviera any day.

All in all, I thought the paper was well received, and after paying a visit to the Cristos monument, I returned home, I was surprised to find that the company had decided not to pay my stipend, because they felt that I had not presented anything "new." Happily, I had retained a copy of their original contract and convinced them that they were obliged to follow through on their commitment. Lesson Learned: you just never know what kind of chicanery you may run into.

The Titan IV "Should Cost" Study at Martin Marietta

In the intervening years there were, happily, numerous small consulting contracts and invitations to give talks in Brazil, Honduras, Costa Rica, Seoul and Berne, Switzerland. There were some interesting assignments, such as a "should-cost" study for the US Air Force on the *Titan IV* missile at Martin Marietta's Littleton fab in the foothills outside Denver, Colorado in February of 1989. Although it afforded us an opportunity to do some weekend skiing in those awesome mountains at Breckenridge and Copper Mountain, I felt frustrated in that my recommendations were readily accepted by the staff, they couldn't get the funds to implement them.

Titan IV Should Cost—Quality Assurance Recommendation 1

Based upon my observations of the Titan IV weld-prep and welding area on the 2nd floor of the factory, I believe that much of the dedicated effort by QEs and the welding staff & management to try to keep materials to be welded as clean as possible, is an exercise in futility. Steve Mullen, QE Dept. and Jack Huck, Major Weld Ops Mgr., have both issued extensive guidance to the weld prep and welding staff on keeping parts clean prior to and during the weld process.

However, the truck and pedestrian entrance at the West end of the building lets in dirt, humidity and debris, all of which will have a deleterious effect on Huck & Mullen's efforts. The seriousness of this condition cannot be minimized, because it has been clearly demonstrated throughout industry that clean parts surfaces are essential to producing a good weld at parts interfaces.

I therefore recommend that an air lock or, at least flexible curtains be installed at the West end 2nd floor factory truck and pedestrian entrance to try to keep out dirt from the weld area.

John Larry Baer, P.E.
Titan IV QA Should Cost Team

That was my diplomatic recommendation. What I had observed was that four sectors of the huge missile body, some 20 feet tall and very carefully machined to assure clean edges, were lined up around a core. They were then robotically welded from top to bottom and then inspected with black light and X-ray to assure a flawless weld, needed to assure structural integrity during missile launch. This process, we were told, cost about a million dollars.

The problem was, that as the welding was taking place, we would periodically hear a ping, indicative of a void or inclusion or dirt. Those faults had to be carefully ground out and re-welded and re-inspected, sometimes twice or three times–at considerable cost. Now during the weld process, as noted above, fine grained dust and sand could easily enter the site. But the modest cost, maybe some $100,000, would have required facility funds, and everyone agreed that it was too much trouble. So we kept wasting millions with this mindless dirt-prone process.

As to the skiing, as we had expected, the conditions were fabulous. We'd get off the lift, trudge up some distance to the coffee shop, and then head down the long, carefully groomed slopes. But then came the end of the day and as I slid into the end of the lift line at the foot of the mountain, I skidded, and realized that I was not fully in control. And to me, whether skiing, or horseback riding, being in control was critical. So that was my last run of skiing–no regrets.

Charge to Ballistic Missile Defense Program Managers (Banquet) 11/3/81

1. The cooks, waiters and other staff personnel were able to set a banquet before you within a cost target, in a timely fashion, hot and tasty. Had the meal merely been hot, or even hot and tasty, but over cost and after keeping you waiting an hour there would have been a lot of unhappy people in this room. .

Your problem as managers is not dissimilar. If your system is produced to specs but is late and over budget, or worse yet, has to be accepted on waivers, you haven't done your job. If you have accepted a design from your designers, which works in DT-I but is gold plated and hogged out of solid bar stock you haven't looked at the producibility.

If you've wisely programmed money to look at the producibility of your system during AD and ED but diverted the PEP funds because you had costly technical problems to resolve, you've just pushed your problems on to the shoulders of the next PM to follow in your post.

If you've decided to produce a system that works but haven't done a cost driver analysis to see where you can bring the cost down, you haven't done your job.

If you use new technology components in your system which promises to give us a quantum improvement in effectiveness–Great! But have you considered if the manufacturing technology exists to make the item at a reasonable cost and in time to feed into the total assembly schedule?

2. We are constantly assaulted with cold, tasteless fare, served late and well above the price we bargained for-not in terms of what we eat, but in terms of the projects paraded by us. And there's always a slew of reasons of why the project is late and over cost and often doesn't work the way it should. My old, field first sergeant always told me: "Excuses and Alibis Represent Failure."

Well, the best way I know of to prevent those failures is to look at the total system and the total production process-top down; tear it apart, step by step, and then put it back together methodically, with someone who's had shop experience and dirt under his nails, to make sure it's put together right and the raw materials flow into your factory at one end, get processed and come out as an inspected, finished product at the other end. Just remember that every time you sit down to a hot meal, well served-if your product isn't hot and on time and within cost–you haven't done your job.

But that was history, when I was still managing Manufacturing Methods and Technology for the US Army. Ahead of me lay not only consulting, but getting involved with CALS!

Part 7
The CALS Program
Computer Aided Acquisition Logistics Support around the Globe

First CALS DoD/Industry/NIST Conference April 1989, Gaithersburg, MD

The first CALS Conference was actually held in three places. It was held on 20 April in conjunction with the National Computer Graphics Association's NCGA Integrate '89 in Philadelphia, Pennsylvania; at the Federal Computer Conference and Defense & Government Computer Graphics Conference FCC/DGC West in Anaheim, California, on 27 April 1989; and at the National Institute of Standards and Technology (NIST) on 2 May 1989. It set forth noble goals.

Computer-aided Acquisition and Logistic Support (CALS) was a DOD and Industry initiative. CALS was to address the integration and use of automated digital technical information for weapon system design, manufacture, and support. And, it was proposed that CALS focus on the generation, access, maintenance, and distribution of the technical data associated with weapon systems. This was to include engineering drawings, product definition and logistic support analysis data, technical manuals, training materials, technical plans and reports, and operational feedback data. It was hoped that the CALS program would facilitate data exchange and access, and reduce the duplication of the data preparation effort. CALS was to provide both the initiative and framework to integrate the existing islands of automation within DoD and Industry. That was the bottom line!

The theme for that first conference was: "CALS–Catalyst for Global Competitiveness." At least that's what my VHS tape recording tells me. For 10 years CALS actually was a major driver of productivity, not just in the U.S., but also in Europe and Asia. International CALS conferences were held in Berlin in 1993, in Seoul and in Athens in 1996, in Frankfurt, in Paris, and in Istanbul in 1997, and once more in Paris in 1998, and I had the good fortune to be part of them all.

CALS Expo '92, San Diego Convention Center, December 1992

In real estate the saying is: "Location, Location, Location." Well, the same holds true in arranging for a conference. The opening conference, held at NIST, the old Bureau of Standards in Gaithersburg, MD provided a distinguished venue that would lend the program credibility. The 1992 conference in San Diego had cachet in that it's a place that people enjoy visiting.

The hotels on Coronado Island with their magnificent views and the great Conference Hall meant that people would want to attend. Future years would follow suit. The tracks on Automated Publishing, Concurrent Engineering, Product Life Cycle Support, Enterprise Architecture, and Information Technology, as well as Product Data Exchange Standards, helped from an intellectual point of view. Another major draw was the list of keynote speakers, such as LTG John Bruen, USA (Ret.,) Lord Chalfont (UK,) M. Henri Martre (FR) (*Chairman, Aerospatiale,*) and Sydney Gillibrand (UK) (*Vice Chair British Aerospace*, and Air Marshall Sir Michael Alcock *(RAF Chief of Logistics Support,)* helped.

The speakers comprised Generals, Admirals and leaders of industry–in other words, it was the kind of conference where you expected to come away with something useful. I still have copies of the extraordinarily elaborate and, at that time quite unique, impressive color, Powerpoint slides used by one of the leading Japanese speakers. The essence of all the CALS conferences this year and in the years to come was "innovation."

CALS Europe Berlin, September 1993

Granted, while San Diego was a tough act to follow, Berlin was hardly second rate. Naturally, since this was the first European CALS conference, the program committee and speakers represented France, Germany, Britain, Denmark, Italy, Netherlands, Norway, Sweden and included a speaker from Japan. The range of topics was broad–as the French say: *Plus ça chose, plus çe la même,* or "the more things change, the more they remain the same." Nobody fell asleep.

The opening talk by *MG Schmidt-Petri,* from the NATO Maintenance and Supply Agency in *Capellen, Luxembourg,* set the tone when he noted that, while CALS Europe was different from US CALS, they were complementary rather than in conflict and highlighted the need for a competitive edge.

To give you an idea of the stature accorded the conferences, one of the keynoters was a member of the Board of *Rheinmetall, GmbH*; another was the Director-General of the Commission of European Communities, and one was the NATO Assistant Secretary General for Defence Support. Not too shabby.

As an interesting side light, Elaine Litman, the Deputy Defense Executive of the American CALS office, and I, decided to visit the famous *Staatsoper* to see one of my favorite operas, Jacques Offenbach's *"Hoffmann's Erzählungen"* (Tales of Hoffmann.) Unfortunately, the staging and costumes were so *avant garde* as to be grotesque. The only way to enjoy the music and the voices, was to close your eyes and just listen. Too bad.

There was, of course, a CALS Conference and Expo in the States every year–in Atlanta, Georgia in 1993. At CALS Expo '91 there were

2370 registered participants, more than 150 speakers (plenary sessions and 12 tracks for technical. sessions), and 140 exhibitors.

CALS EXPO '94 was held in Long Beach, CA from 5-8 December. And there was CALS '95, and CALS/ISG (Integrated Logistics Support) Summer 1996 Symposium, Annapolis, MD, and in August 1996 the National Materials Advisory Board, Panel on Explosives Detection Systems

Meantime, there were *ilce* conferences, which we helped to arrange and provide distinguished speakers from around the globe, not just the U.S. There was ILCE '94–Integrated Logistics & Concurrent Engineering, held in Montpellier, France, *Fevrier* 1994, ILCE '' '95–Concurrent Engineering & Technical Information Processing, at *Espace Saint-Jacques, Paris, Fevrier 1995, and* ILCE '96, the 4th International Conference on "Information Systems, Logistics ntegration, Concurrent Engineering and Electronic Commerce," at *Hôtel Mercure, Paris, Octobre 1996*

Beside the *ilce* conferences, which we helped to arrange and provide distinguished speakers from around the globe, not just the U.S. there were great CALS Conferences every year in very desirable locations, e.g.

Korean Management Advisory Council Conference 29 April 1996

I had been invited by the Korean Management Advisory Consultants to

give a briefing in Seoul on *"Concurrent Product & Process Development" or "Concurrent Engineering."* Well, having had excellent advice from experts like Naomi McAfee and others, this was going to be a piece of cake. I had prepared an extensive set of Vu-Graphs on the benefits of *Electronic Data Interchange* and, in particular, citing the major corporations around the globe who were implementing it, as well as what Americans thought were important aspects of their careers.

CALS Pacific '96–Seoul, Korea, 3-6 September 1996

"Challenge for the 21st Century"

For the second time in one year I was invited to come to Korea to give a presentation–this time for the CALS Pacific conference. I put my slides together and with tickets in hand, provided by my hosts, set off once more for the tedious flight to Seoul. To make sure that my briefing was as up-to-date as possible, I reviewed the slides I had on my laptop, during the flight–and quickly realized that I needed two introductory slides.

As soon as I got to my room at the hotel, I called down for a converter to plug in my laptop, added the two introductory slides, closed the laptop, and went peacefully to sleep. Next morning my hosts asked me if I would mind speaking in the morning, rather than the afternoon, as previously scheduled. Of course, I agreed and proceeded into the conference room. Then the projectionist asked if my laptop was set to project 9' x 12' or whatever, and he said that I needed to turn off the laptop and reprogram it for the 12' x 14' screen. No problem, except that–oops–I had forgotten to save the two new slides I had created the night before. So, even as people started entering the hall, I recreated the two introductory slides right there on the podium, in time to start my talk. Happily, it all went well.

12th International Logistics Conference, Athens/ Glyfada, Greece, 27-29 September 1996

What was special about this conference, was that I was invited to present a 4-hour "CALS Tutorial" in *Glyfada*, a suburb of Athens. There were 245 attendees from 22 countries, including Australia, China, Finland, Israel, Japan, Russia, Singapore, South Africa, and Yugoslavia, in addition to almost every country from Europe and 143 from Greece. I had already given a talk on *"Pre-planned Logistics and Maintenance for Disaster Avoidance"* in the morning.

When I started my tutorial after lunch with about 25 in the audience, I expected that some would doze off, and when I gave them a smoke break, I fully expected that I would lose a few listeners. I was wrong– for the second hour my class grew to 35, and for the 3rd hour I had 45! That felt good and gave me a lot of satisfaction. I had also arranged for the keynote speaker, Dr. Sarah James, president of the Society of Logistics Engineering, and the track coor-dinator for the *Systems Logistics* track and several of speakers, including Don Keith of SAIC.

Global Electronic Commerce & Money Management, Athens, Je '98

Since my seminar at the 12th ILC in Athens in 1996 had gone so well, my friend, Leo Lambert, the head of SOLE in Greece, (since deceased) invited me back to give another seminar to the *Federation of Hellenic Information Technology Enterprises* about "Global Electronic Commerce & Money Management. I have two mementos from the occasion:

On the left was my opening slide in Powerpoint, that was the popular means of providing information in those days and lent itself to storage on a laptop computer that one could drag to whatever site was needed. On the right is a reminder I've kept on my desk ever since that ominous evening when I was invited to also address the Athens chapter of SOLE. I had not realized that I would be preceded by a Government Minister, who spoke at great length in Greek, of which I did not understand a word. The picture was to remind forever after NOT to ever show my feelings so openly again. The invitation from Athens read as follows:

Date: Wed, 20 May 1998 12:58:22 +0300. Subject: SOLE Seminar
From: Harry Angelopulos <unicon@acropolis.net>
To: "BAER, JOHN" <jbaer@capaccess.org>
This morning I spoke with the cognizant people at SOLE regarding the seminar, which we tentatively plan for Monday evening. The topic will be "Quality in Government Procurement". The idea is for the

various Greek speakers to present the existing procedures for working with the Greek Government, to present the applicable regulations and to present areas for improvement. It was suggested that you present (with the aid of viewgraphs) some ideas and methods about improving the procurement procedures.

In other words they would be interested in hearing your input regarding how the US and several other countries have implemented EC & CALS to improve the way they cooperate with their suppliers. Please review the above and let me have your comments & suggestions
  Regards, Harry

When I finally got to speak, the talk was actually well received. Greeks, like folks in Spain and Italy, are used to staying up late, and so I had no one falling asleep during my talk.

Two key elements of my talk were a chart showing how the banking industry led in adopting the *Ethernet* to enhance business, and my "charge" of the need for:

Ownership vs Authorship
&
Commitment vs Involvement.

I believe that the age-old analogy I used to define this, was, that in making bacon and eggs, the chicken is involved, while the pig is committed. Besides establishing contacts with leaders from the Hellenic Aerospace Industry, the Hellenic Navy, and the Chambers of Commerce & Industry of Athens, Iraklio, Piraeus, and Thessaloniki, I got the chance to visit *Hellenic Arms,* and to learn about their casting of non-ferrous metals for small and large caliber projectiles and cartridge cases. That meeting resulted in the following letter:

Mr. Nick Angelakis, Deputy Planning Director
Hellenic Arms Industry SA   nickangelo.ath@forthnet.gr
160, Kifissias Avenue   tel. 011-30-1-674-2611
115 25 Athens, Greece   FAX: 011-30-1-674-2715
29 June 1998
Dear Mr. Angelakis:

At the SOLE meeting you reiterated the interest you expressed at the Money Show, for ideas on using CALS in procurement planning and ERP or Enterprise Resource Planning. My off-hand recommendation would be that, as a first step, you must have the support and understanding of your top management to harness the CALS tools. Without that you are guaranteed failure.

Unfortunately, implementing CALS is not cheap. Scanning drawings into the computer, capturing manufacturing methods and technology that now exist only in some workman's head, and recording all the materials, names, clients, and customers into the computer for rapid future access, all take time and energy, and money. BUT-once entered, it makes life so much easier that you wonder how you ever got along without CALS.

The other important step in procurement planning and ERP is to identify and work closely with your client(s). It does no good to be able to use EDI (Electronic Data Interchange) in-house & with some of your suppliers, if your client can't access the database. You will also need to identify all the STANDARDS you now use and what is Best Practice in industry (Greek, European & American.)

Finally, before trying to apply CALS tools to create an agile enterprise within Hellenic Arms, it is first necessary to remove non-value-added steps from all operational procedures, both in the office and on the factory floor. Otherwise you will just be speeding up a bad process.

Here is in the U.S. we spent many millions in just straightening out the production lines and parts storage facilities at places like Watervliet and Rock Island Arsenals and related Depots. BUT-once implemented, these simplified operational procedures exceeded all expectations in the speed and quality of the production process.

"General Electric Co. bought $1 billion worth of supplies via the InterNet last year. That saved the company 20% on materials costs because its divisions were able to reach a wider base of suppliers to hammer out better deals. By 2000, GE expects to be buying $5 billion over the Net." (according to BUSINESS WEEK's Information Technology Annual Report, dated 22 June 1998, pg. 125.) "At Boeing Co., there are 75 projects for using the Net to connect to contractors and customers-everything from zapping documents to the government, to tracking the history of every plane Boeing sells." (ibid) This is just a short list of potentially profitable actions. Please let me know if I can be of further assistance.

sincerely, John Larry Baer, P.E., President, IMEC; Home Page: http://netvoyager.com/IMEC

4206 Elizabeth Lane, Annandale, VA 22003-3652, USA

Tel.1-703-323-6952; FAX: 1-703-323-9045; jbaer@capaccess.org

The reason I included this letter is to reflect the very nice relationship we, as a consulting firm, had with foreign clients, who were anxious to learn the latest technologies, which at that time, originated in the U.S.

CALS Expo '96, Long Beach Convention Center, Long Beach, CA October '96

CALS Expo '96 was probably the high point of CALS conferences, with

some 3200 attendees, including 2044 American and 840 foreign visitors. The largest contingent, 366 from Japan, accounting for 43% of the foreign visitors, followed by 176 from the UK and 55 from Canada. There were even 5 from Russia.

CALS Expo '95 had also been held at Long Beach, but it had attracted 2055 US and 745 non-US conference attendees, including 412 from Japan, 103 from Korea, 62 from the UK and 52 from Canada. Attendance had actually peaked in 1992, and so, over the years it became obvious that folks were changing their allegiance and interest to other subjects.

Mediterranean CALS '97, Istanbul, Turkey

"Continuous Acquisition and Life Cycle Support Commerce at Light Speed"
(Conference Chair and Coordinator) 11-13 June 1997
Then there was the invitation to arrange speakers for the Mediterranean CALS '97 conference in Istanbul, Turkey (where I wound up chairing the conference) and to do some great sightseeing afterward, including the *Fairy Chimneys at Pamukale*. I had been meeting and in correspondence with Major *MustafaErcin* (who asked to be called *Bimbashem)* and Colonel Aksu of the Turkish Army, to help them arrange for speakers and exhibitors from around the globe, for this *Turkic* CALS conference and had gotten a nice lineup for them.

Among the speakers I had managed to line up to speak at the conference, were:
Professor Ben Blanchard, Virginia Polytechnic Institute
Michael Conroy, DESCO, Paris
James Craft, SRA International & David Chesebrough, IRIS LLC, Fairfax
Dr. Marc D'Alleyrand, President of Info & Tech Transfer, France
Mike Daniels, Director of Configuration Management at CACI
John Downes, Procurement Executive of the British Ministry of Defence
Ali Kutay, CEO, Formtek, Pittsburgh
Mrs. Elaine Litman, DoD Exchange Execuitve at Lockheed Martin
Professor Kulwant Pawar, University of Nottingham, England
Brent Pope, Senior Executive at Coopers & Lybrand
Philippe Thomas, Dassault Aviation, St. Cloud, France
Wan-Jae Yu, DAEWOO Information Systems, Seoul, Korea.

On the day before the conference, June 10th 1997, we had arranged for a CALS Overview tutorial by Professor Tom Gulledge of George Mason University, and in the afternoon I gave a 3-hour tutorial on

Commercial CALS and Electronic Commerce. Thank goodness, Powerpoint slides existed even in those days, making the presentation far easier than reading from a script.

The next day, Brigadier General *Cemal Alagöz*, opened the conference in a 1½ hour talk, followed by the Turkish Minister of Defence, who held forth for another hour–in Turkish, of course. *Bimbashem* had disappeared from the dais, along with the General, and I found myself alone on the dais, except for the speakers I brought up in turn. At the coffee break when I asked him how come I was left there all alone at the *dais*, he said casually: "oh yes. You're the chairman. Didn't you know?" Ah well, the conference went well and after it was over, we went in our rented car and driver to visit the fantastic sites at *Cappadocia, Guereme, Izmir, Ephesus, Kusadasi and Pamukkale.*

All of the travel arrangements had been made by VIP Tourism, a local travel agency in Istanbul, who underwrote the conference, expecting that it would result in some payback. Unfortunately, all the Army attendees were admitted free of charge, so that, even with scrimping by putting us up at a very basic two-star hotel, they lost money. In fact the *Taxim Plaza* Hotel was so basic, and so poorly located that some of us opted to upgrade to the Hyatt Hotel, which was adjacent to the Military Museum Cultural Center, the site of the conference (at our own expense, of course.)

Fortunately the agency was able to arrange a bus tour of Istanbul for us, to include a visit to the spectacular *Süleymaniye Mosque,* with its four slender minarets and huge, courtyard, and the world-famous *Sultanahmet* or "blue" Mosque. We also got to tour the Basilica of St. Sophia, now called the *Ayasofya* Museum. They also took us on a walking tour of the covered Bazaar–a shopper's delight, where I was pursued by a souvenir vendor, who finally slapped a tourist map against my chest–and deftly lifted out the travelers' checks I had in my shirt pocket.

Since the travel agency lost so much money underwriting air-fare and hotels for all the speakers, they were unable to pay for our sightseeing, except for a Bosphorus cruise, which was wonderful, but it was all well worth the expense to see these extraordinary places. The boat went under the awesome and beautiful bridge, that links the European side of Turkey with the Asian side. That bridge spans the Bosphorus strait, and it's located between Ortaköy Mosque on the European side, and Beylerbeyi on the Asian side. At 1074 meters in length, this beautiful suspension bridge is the 13th longest in the world. When it was completed in 1973, it was the 4th longest suspension bridge in the world and the longest outside the United States of America.

On the day after the conference we had a car drive us through (or rather, around) Ankara and on to *Cappadocia*. We even got a chance to stretch and have lunch at a typical *caravanseri*–a sort of Turkish inn, where caravans stopped with their camels on their trade route.

After a very comfortable overnight stay, a very heavy-handed Turkish massage and a delicious dinner at the modern *Dedeman* Hotel, we had our first treat with a visit to the *Kaymakl Underground City*. The houses in the village are constructed around nearly one hundred tunnels of the underground city and are still used today as storage areas, stables, and cellars.

Four floors are open to tourists, with each space organized around ventilation shafts. The rooms are all of different shapes, because the design of each room or open space is dependent on the availability of ventilation. What was a bit scary was that one had to squeeze through a rather tight space to get into the underground city. It was easy to understand when our guide told us that some time ago a rather corpulent visitor got stuck in the opening and was pried out only with great difficulty.

The afternoon was spent walking around the hills of *Göreme*. We learned that after the eruption of *Mount Erciyes* about 2000 years ago, the lava that formed soft rocks in the region, covered a region of about 20,000 km^2. The softer rock was eroded away by wind and water, leaving the hard cap rock on top of pillars, forming the present-day *fairy chimneys*. People of Göreme, at the heart of the Cappadocia Region, realized that these soft rocks could be easily carved out to form houses, churches, monasteries–providing fantastic tourist sites for us.

The next day we flew to *Izmir* on the *Aegean Sea* and then drove to the *Korumar Hotel in Kusadasi*. There, sitting by the Aegean Sea as we had our meals, we had a chance to reminisce about those fantastic fairy chimneys, and make friends with other tourists.

It was our setting-off point to visit Ephesus, an important center for early Christianity, and later, the capital of proconsular Asia, which covered the western part of Asia Minor. The original city of Ephesus was located on low ground, and was completely flooded by the sea. The city was rebuilt in 292 BC and the inhabitants relocated to the new city that exists today. Many spectacular facades remain, such as the *Library of Celsius,* whose second floor had a passageway that connected to the brothel across the way. *Ephesus* also had an imposing public toilet, where seats were located around the room, making this a gathering place for conversation, as well as other functions. In typical Roman fashion, it had an excellent piping system, and the public baths next door offered a *tepidarium* with a hot water pool, as well as a *caldarium* for a refreshing cooling dip.

The Roman amphitheater with superb accoustics, could accommodate 25,000 of the city's 500,000 inhabitants. The city had one of the most advanced K"http://en.wikipedia.org/wiki/Aqueduct"\o"Aqueduct"aqueduct systems in the ancient world, with multiple aqueducts of various sizes to supply different areas of the city, including 4 major aqueducts. Engineers take note! We also got to see what was purported to be the modest stone structure, where Mary was said to live out her days, with occasional visits from the apostle, Paul, who made his base there.

In passing, it should be noted that the loss of its harbor caused Ephesus to lose its access to the Aegean Sea, which was important for trade. When the Seljuk Turks conquered it in 1090, it was a small village. The Byzantines resumed control in 1100 and kept control of the region until the end of the 13th century. After a short period of flourishing under the new rulers, it was abandoned in the 15th century and lost her former glory. Today, the harbour is 5 km inland.

CALS Europe '97 Frankfurt-am-Main, Germany 1-2 *Oktober 1997*

CALS & Electronic Commerce

OK–back to work. 1997 was a very busy year–a year in which, beside Turkey, I got a chance to go to Germany, Brazil, (Orlando, Florida too) and France. I had been invited to speak at *Euro CALS* on "The Payback from Electronic Commerce." The only problem was that the date happened to fall on the Eve of Rosh Hashonoh, the Jewish New Year.

What was I to do? Well, the committee was nice enough to schedule my talk for the morning, so that after speaking, I was able to attend services at my old synagogue on *Freiherr von Steinstraße.*

The temple had been well restored after the destruction wrought by the Nazis, and it filled up rapidly, with everyone talking up a storm– even as the Rabbi extended his greetings and the cantor began to chant. When I couldn't stand it anymore, I turned to the people behind me (who were chattering in Russian and other Slavic tongues) and told them in my best Frankfurter German dialect: *Schlappmaul. Halt die Gosch.* (blabbermouth–shut up already.) They looked at me like a foreigner–and kept right on talking. I left my old *Shul* in tears and went to my hotel room to pray. Happily, the next morning I was able to fly home and attend evening services at our temple.

21st Century Commerce & CALS EXPO USA 1997

Orange County Convention Center, Orlando, Florida 13-16 October 1997

By 1997 the center of the CALS universe was obviously shifting from the USA to OCONUS venues. Where in 1996 there were 838 foreign attendees, in 1997 there were only 324, conference registration was down from 1200 to 750 and even the number of exhibitors was down from 1144 to 934. That was about the last of the events labeled as CALS. DoD abandoned it about then, although ADPA/NSIA tried to keep the effort going with their AFEI (Association for Enterprise Integration) group. The following letter and table from GIC France seems to verify my conclusion here:

GIC France Guidelines for a new European CALS/ ELECTRONIC COMMERCE

While CALS and Electronic Commerce concepts and practices are spreading all around the world, at least within administrations and major industrial companies, the number of events (conferences and/or exhibitions) organized on themes related to these concepts is also increasing rapidly. As these events are soliciting the same limited kernel of speakers and exhibitors and the same potential attendees, the result is, in several countries, a decreasing level of interest of the conferences (the same speaker is presenting the same speech to different events) and a decreasing number of exhibitors and attendees.

For example, the following events were organized in Europe in year 96 (list not exhaustive):

• Belgium: CE Technical Day : Brussels, June 26

• France : MICAD : Paris, March, SGDT'96 : Paris, May, CALS Europe'96 : Paris, May 28-31

EDI '96 : Paris, June, Engineering Systems Design & Analysis : Montpellier, July 1-4,

ILCE '96 : Paris, October 14-17; GITI/ESoCE Workshop on Concurrent Engineering :

Paris, October 17 (in conjunction with ILCE'96); Electronic Commerce : Paris, December

• Greece : 12th International Logistics Congress : Athens, September 27-28

• Hungary : 10th European Simulation Conference : Budapest, June 2-6

• Ireland: Conference on Integration in Manufacturing 1996 : Galway, October 2-4

• Italy :   8th European Simulation Symposium : Genoa, October 24-26,
ICE '96 Workshop : Milano, November 7-8
• Switzerland :   Product Knowledge Sharing : Basel, October 30-31
• United Kingdom :   APLS'96 : London, April 16-18
In addition some other events may also compete with the previous ones, these are for instance in Paris: LE BOURGET : in June, each odd year; EUROSATORY : in June, each even year (June 24-29, 1996); EURONAVAL : in October, each even year (October 21-25, 1996)
Yours sincerely,
Pierre-Yves VILCOQ, *Groupement des Industries pour CALS en France 22/11/96*

I don't know why I feel the need to justify myself, but just to verify my credentials, two letters I wrote on behalf of the CALS steering committee, are shown in the center of this book:

Happily, my associates in France also invited me to arrange for speakers to: ILCE 97–*Enterprise Intégrée & Commerce Electronique, Paris, 25-26 Novembre 1997*

CALS Europe '98 & ILCE '98, *Palais des Congrès, Paris, France,* 16-18 September 1998, and for IMEC to make a presentation to SOLE Europe on EMSYS (Engineering & Maintenance Systems) and TIMS (Technical Information Management System in Frankfurt in Sept. 1998.

IV Simposium Industrial "Ingeniera de Vanguardia"

Tegucigalpa, Honduras, Febrero 1999
So now, that my professional career seems to be coming to a close, there were still a few very pleasant opportunities to reach out with what I had learned over the years and pass it on to others. I had been invited to deliver a talk at an Industrial Engineers Conference at the Universidad Tecnologia Centroamerica in *Tegucigalpa, Honduras* on 6 November 1998. Unfortunately, on 30 October 1998, the city was

heavily damaged by Hurricane Mitch, which destroyed part of the city, including 10 of the 12 bridges that connect Tegucigalpa and *Comayagüela,* as well as other places along the riverbanks of the *Choluteca* river. The 3-story education Ministry was flooded to its top floor, destroying all of its records.

The hurricane remained over Honduran territory for five days, dumping heavy rainfall late in the rainy season. The ground was already saturated and could not absorb the heavy precipitation, while deforestation and debris left by the hurricane led to catastrophic flooding throughout widespread regions of the country, especially in Tegucigalpa. The cities of Tegucigalpa and Comayagüela, jointly, constitute the Capital of the Republic.

I had been invited by the young (24 year old) dean of Engineering at the University in Honduras, to deliver a talk on *Comercio Electrónico y Logistica Usados Satisfactoriamente en la Ingeniería Industrial"* (Electronic Commerce & Logistics used in Industrial Engineering.) A report of my findings may be of interest to the reader.

Report on the Sights of Tegucigalpa, Honduras

As an invitee to speak at UNITEC's Industrial Engineering conference, I thought I should take the opportunity to see how I might be able to help Honduran companies devastated by Hurricane Mitch. In addition I wanted to see if we could help Honduran industry to set up a facility to fabricate low cost building panels to quickly replace the homes of people left homeless in the extremely flood-prone Rio Grande riverbed.

With the help of a young UNITEC I.E. graduate and the Director of ANDI (*Asociacion Nacional de Industriales de Honduras*), we visited the very small firm of *Closet y Puertas* (cabinets and doors) where they fabricate modest metal cabinet doors and some wooden kitchen cabinets using salvaged and home-made sheet metal equipment. On a far more profound scale, we visited *Fabrica de Cajas de Madera* where men were trying to locate and then, using shovels, dig out totally mud immersed pieces of equipment, ovens, and even file cabinets of ravaged, water and mud damaged paper files. In one corner, on top of the 3-meter layer of mud, the river had dumped a truck-through the roof of the factory-see photo on page 121.

We also visited Textiles Rio Lindo, where they are completely rebuilding their water treatment facility to be able to use the sewage-laden water from the Rio Grande. Our final visit was to a Honduran wood industry firm run by a dynamic young UNITEC graduate.

The region's major problem in my eyes is the restoration of bridges and roads. With nine of 12 bridges connecting Tegucigalpa and its sister city, *Comayaguela* across the Rio Grande and Rio Piceno, still standing, traffic at the remaining bridges and ALL roads leading to them, is a constant, nerve wracking hassle. Many of the roads leading to nearby factories are deeply rutted, with deep pot holes and shoulder often totally washed out-making driving extremely hazardous.

Every help must be extended by the American and European engineering world to help Honduran factories to rebuild and provide good, used equipment they need to rebuild at the lowest cost and fastest pace. They have the will, but they need the means-and that includes information on recycling precious resources like their water, and how their workers could work more productively and safely.

Honduras' potential extends from the dedicated folks who own and run the factories to their most outstanding asset-their bright, intelligent, young people. The students and young graduates I met at the *Universidad Tecnologica Centro-Americana* (UNITEC) represent the greatest resource for this devastated country.

But–to start at the beginning-It's not often that one receives a royal welcome, and thus the abrazo extended by my young hosts in the VIP lounge of this little airport, was not only unexpected, but a delightful surprise. After the initial cordialities, UNITEC's very young Dean of Industrial Engineering spotted the widow of the late and distinguished Mayor in the lounge, being interviewed.

This charming lady, who has now stepped into her husband's shoes, became Jessica Calderon's immediate target to introduce this humble traveler and my planned mission to aid in the rebuilding of the Hurricane battered region. Madam Mayor, in typical, gracious Southern charm, invited us to a meeting in her office with her deputy responsible for the reconstruction.

(It should be noted that she has taken up the cudgel her late husband wielded until his untimely death in a helicopter crash while inspecting the hurricane damage-to get a law passed which would preclude people from re-building in the extremely flood-prone Rio Grande riverbed. She is actively continuing his work of erecting modest shelters for the displaced, in a safe place, high on the hill overlooking their former homes.)

From such a moving beginning, I knew this was going to be a great trip, and, at the end of just one day, I feel that the people I've met and the interest stirred up in working together to rapidly build thousands of homes for hurricane displaced people, is already well established. Antonio Fraile, the young *Coordinador de Operaciones para Organizacion Internacional para las Migraciones*, is not interested in constructing a plant for agricultural or wood fibre waste board. But, he wants to know how much the sandwich material with foam insulating core from World Building Systems would cost for 10,000 six by four meter homes.

The ¼ inch wood fibre filled cement boards have provided adequate squatter shacks high up on the hill, and well removed from the Rio Grande level shacks the displaced residents had opted to return to. To see the location of these corrugated sheet and cardboard structures under the bridges and in currently dry, open stretches, is to boggle the mind of anyone who can look just a few months ahead to even modest rains that would likely, swiftly wash away their tarpaper hovels.

The provisioning of elemental housing for these hapless folk is the kind of manageable task that could provide quick results, unlike the Navy Seabees' longer term rebuilding of some of the 10 bridges ravaged and washed away by Hurricane Mitch, leaving intact only 2 bridges to connect Tegucigalpa with its neighboring *Comayaguela*. All these bits of wisdom were fed to me by Blanca Trinidad, a recent UNITEC graduate and charming chauffeur and raconteur, during an extraordinary day of touring and photographically recording the horrendous damage the storm wrought.

In addition, guided by Romel Barahona, Director of ANDI (*Asociacion Nacional de Industriales de Honduras*) and his young, charming logistics aid, we visited the very small firm of Closet y Puertas. Here some eight people, with a dog running between their feet are fabricating modest metal cabinet doors and some wooden kitchen cabinets, in many cases, using equipment they have painfully restored or built after Mitch destroyed his building & most machinery.

This good man, who has the knowledge to do greater things, is trying to keep people usefully occupied with this modest effort, taking sheet metal, measuring, marking, scoring, cutting, bending, forming and joining the pieces to form usable, marketable cabinet doors. To assist him, we need to identify for him second hand breaks, and other sheet metal working equipment he could purchase, so that his people can be more productive than they are with their present manual operations. On a far more profound scale, we visited *Fabrica de Cajas de Madera*.

Here women were salvaging bundles of sticks which had been made for use as ice cream sticks, and men were trying to locate and then, using shovels, dig out totally mud immersed pieces of valuable equipment, ovens, transfer tools and even file cabinets of ravaged, water and mud damaged paper files. In one corner, on top of the 3-meter layer of mud, the river had dumped a truck-through the roof of the factory. While heavy excavating equipment and dozers can be used to some extent, the excavating job is mainly one of pick and shovel, in order to minimize the damage to the mud-submerged tools.

The need here, beside loans to hire back laid-off people for digging out the salvageable equipment, cleaning it and relocating it to a remote site in the hills, where the plant is operating, is for technical information on the safe cleansing and equipment restoration. It would seem that such information should be available from recently flood damaged towns like Rock Island, IL and several of our Southern and South-Western cities which have seen similar damage, but not on as egregious a scale as Hurricane Mitch. None, I'm sure can point to soft, fine beach sand washed 100 miles up-river from the ocean, as we did at this location.

My only other observation and recommendation for US aid, might be the need for greater emphasis on the swift rebuilding of the remaining wash-out bridges, to help restore some degree of sanity to a chaotic, sclerotic traffic scene, no matter where or when you try to move about *Tegucigalpa and Comayaguela* amidst hundreds of horn honking freelance taxicabs.

Visit to Tegucigalpa, Day 2

With my trusty sidekick, Blanca Trinidad at the wheel, we picked up our escort from ANDI, Romel Barahona and his Logistics pro, Marivel Banegas and headed for *Fabrica de Muebles y Puertas* (and, before Mitch, y Ventanas.) Here, Jacobo Arevalo, the owner, and Wilfredo Alvarado, the production manager, escorted us through a quite well organized and run factory making some quite well made and finished cabinets, cribs, bunk beds and doors. The plant here had ONLY 1 meter of mud, plus 2 more meters of water covering the equipment, so that they had a relatively easier time uncovering and restoring their woodworking machinery.

With a resident staff mechanic to oversee this process, they had restored the machines and were basically back in full operations. There were, of course, areas that had been destroyed by the rushing water of the nearby Rio, but the owner and manager had used this as an opportunity to rebuild better, and to expand. Their exhaust system for the spray painting area had been demolished and hence the painter was spray painting by hand without a face mask, despite the manager's periodic reminders to wear a mask. (OSHA would have a fit here.)

When asked how we could help, Sr. Arevalo noted that he would be interested in buying some good, used woodworking machinery such as mortisers, tenoners, planers, routers and a radial arm saw. I would offer only one with a built-in safety guard, as his workman insists on operating without one. When I asked Sr. Alvardo how many accidents result in lost fingers, he noted probably only about two a year-that's two too much.

The afternoon tour to the Rio Lindo Textile plant was the highlight of my trip (so far). Here we're talking about a large enterprise by any standards, with several hundred employees, which has taken Mitch as an opportunity to rebuild the facility almost from the ground up. We were greeted by:

Francisco Castillo, the General Manager.

Gustavo Zelaya, Human Relations Manager (our primary point of contact)

Brenda Dominguez, Sales Manager, Honduras
Rigoberto Alvarado, International Sales Manager (*Gerente de Ventas*)
Carlos Vijil S., Treasurer (*Tesorero*)
and later joined by B.B.N. Laxman Mendis, Ingeniero de Planta, from Sri Lanka.

The plant fabricates 100% cotton, 65% polyester-35% cotton & 65% polyester-35% Rayon from the baled raw material, going through carding, drawing, roving, spinning, winding, warping and threading before proceeding to the looms. There are ample opportunities for the material to break along the way, and the staff expressed great interest in improving quality control at every step, to minimize the need for knotting at subsequent steps.

They are also interested in performing time and motion studies, and for this operation I recommended that they engage some of the industrial engineering students at UNITEC in an apprentice program. I also suggested that they might want to hire someone like Neal Schmeidler of OMNI Engineering, to lay out the procedure, and that they might want to consider using Neal also for the supervision of the students.

Another item of potential interest is that Rio Lindo is getting rid of 450 old, but fully operational looms, capable of producing 650,000 yards per month, including a full supply of spare and repair parts. They could even provide technical support to the potential purchaser. They are in the process of acquiring new looms from a bankrupt firm in South Carolina. In the meantime, rather than laying off the staff, while waiting for the new equipment, they are using them to tear down and totally clean and overhaul old equipment, which they plan to continue to use after the factory overhaul. A tangent benefit of these new looms is that they will operate at about 60-70 db noise level, vs. 90-110 db for their old looms. Francisco Garcia, the general manager, is very interested in obtaining more information about OSHA, and how their procedures might assist Rio Lindo operations.

An item of major concern for Laxman Mendis, the plant engineer, was the inability to adjust temperature and humidity controls independently on their Swiss Luwa controls, and that they can't even turn them off for weekends. They are very interested in any means to achieve these two, seemingly obvious, control capabilities.

The area where materials are dyed, washed, dried and all related steps requiring the use and disposal of chemicals, was of great interest, as we discussed the potential benefits of recycling as much of the water carrying the materials-not just to reduce down-stream pollution but for the potential recovery of expensive materials-much as the Radford Army Ordnance Plant in Radford, Virginia, recovered 2000 pounds of acid by recycling wash and process water.

The last process area visited, where the cloth is actually woven, Laxman Mendis, showed us, with great, and justifiable, pride, the outstandingly smooth floor paint surface they were putting down here, to make sure that none of the cloth touching the floor could snag and tear. The concern here, of course, is the potential for workers slipping on the ultra-smooth floor, and the need for extra caution, and probably special footwear.

The final area visited was the water treatment facility. This part of the plant was totally destroyed by Hurricane Mitch, and Rio Lindo, is building totally new sluice ways, treatment tanks, floculators, etc, to make the completely untreated, sewage infested river water usable for their production process.

In addition, they will consider recycling as much of their process water as possible, in order to recover usable materials, that might otherwise by wasted. Rio Lindo is considering engaging some engineering consulting services in order to enhance their speedy restoration AND to assure their competitiveness against such cheap labor textile producers as exist in Asia and elsewhere. Their greatest need is to first create a well defined time line as to when Rio Lindo expects various parts of the plant to become operational. In addition, Rio Lindo will try to identify areas where they could use outside help in reviewing and enhancing their processes.

Day 3, 4 and 5

After the heartwarming and encouraging visit to Textiles Rio Lindo, one would have thought-things probably won't get much better than this; but they did. On Wednesday we were able, with the gracious help of Luis Zelaya, UNITEC's young Dean of Engineering, to reach into the uppermost managerial level of the Honduran wood industry. Sr. Guillermo Bennaton Ramos kindly agreed to meet with us to discuss the potential for making panels for Honduran housing from wood scrap (OSB or Oriented Strand Board) and ASB or Agricultural Waste Structure Boards. To make life easy he offered to send a car to pick us up.

At the appointed time, a young man drove up in a carry-all. Modest, simply dressed, he introduced himself as Eduardo. Gradually we learned that our "driver" was the young president of HonduForest, Eduardo Benneton Ramos and the son of Guillermo Benneton. With great pride, he showed us the samples which he had created from pressed wood, including well designed support columns and bases, beautifully finished doors with pressed wood cores, and other items that would be less expensive than those constructed of solid wood. He was extremely interested in the material we brought him about using scrap wood agricultural waste boards. In fact, he offered to make samples to show us, using the same type of foam core.

In fact, the next afternoon he delivered to the conference site samples of 1/4" blockboard faced foam, using two 1" thick foam panels. He was in the process of pressing a similar piece with 1/4" bloackboard facing, and estimated that he could readily make such 4' x 8' panels for a cost of 1000 *Lempiras*, or about $58. He thought, however, that the panels could be made even cheaper using a local, fast growing *Espinas* or Carbon wood, which the farmers consider to be a nuisance. It is cheaper and much harder than pine and would probably cost less than ASB or OSB, and would not required the composer compression equipment. *Espinas* grows to only about 2 to 3 meters in height, but their supply at present is good for about 10 years. Eduardo, one of UNITEC's top graduates, offered to send further details about this option, but he is still interested in the ASB and expects to sign a letter of agreement with World Building Systems.

It is worth noting that we also met on Wednesday with Sr. Rafael Vallardes, the planning chief for Mayor Dra. Vilma R. de Castellanos, widow of the late mayor. Sr. Vallardes expressed interest in the ideas we offered about providing low cost material for rapid erection of the many thousands of shelters the area needs to locate its displaced citizens, who, much to the Mayor's chagrin, have built shacks right back in the river bed where their previous homes were washed away by Hurricane Mitch. Needless to say, the Mayor and her staff consider housing in safer locations of the thousands of displaced citizens one of their highest priorities.

We also met with Daniel Peters, 3rd Secretary for Economic Affairs at the U.S. Embassy and Rossana Lobo, with the Commercial Service at the Embassy. They are both aware of the need to re-house the Hurricane displaced people, and expressed some interest in the concepts for building a factory to make the OSB and ASB panels and also for the interior, waterproof WEDI panelling from Germany which we showed them.

CONCLUSIONS & RECOMMENDATIONS

The city of Tegucigalpa has three problems: 1) finding property they can buy to locate the housing; 2) materials and designs for quick erection of modest but sturdy and safe housing; 3) financing. The region's major problem in my eyes is the restoration of bridges and roads. With nine of 12 bridges connecting Tegucigalpa and its sister city, Comayaguela across the Rio Grande and Rio Piceno, still standing, traffic at the remaining bridges and ALL roads leading to them, is a constant, nerve wracking hastle. Many of the roads leading to nearby factories are deeply rutted, with deep pot holes and shoulder often totally washed out-making driving extremely hazardous.

Finally, every help must be extended by the American and European engineering world to help Honduran factories to rebuild and provide good, used equipment they need to rebuild at the lowest cost and fastest pace. They have the will, but they need the means-and that includes information on recycling precious resources like their water, and how their workers could work more productively and safely. Honduras' potential extends from the dedicated folks who own and run the factories to their most outstanding asset-their bright, intelligent, young people. The students and young graduates I met at *Universidad Tecnologica Centro Americana* (UNITEC) and at their Industrial Engineering Conference, from Jessica Calderon, their very young Dean, to the youngest freshman, were the brightest and most dynamic people and the most interested listeners I have had the pleasure of meeting in a very long time. They represent the greatest resource for this devastated country.

The heavy rain caused flash floods of the Choluteca river's tributaries, and the swollen river overflowed its banks, tearing down entire neighborhoods and bridges across the ravaged city. The rainfall also triggered massive landslides around El Berrinche hill, close to the downtown area. Needless to say, my talk had to be postponed till February 1999, but even by then, most of the devastation was still visible everywhere we went, as shown on page 121 of the picture section, including the truck that was swept through the roof and boiler inundated in mud in Tegucigalpa factory

I had had pretty good relations with clients and associates in the past, but I was not prepared for the warmth of the welcome I received at UNITEC, where the young (27-year old) Dean of Industrial Engineering, *Jessica Calderon,* proceeded to call me *tio Juan* (uncle John) and addressed me as such in correspondence. (It was a little bit like Moshe Dayan being called *Dodi* (uncle) *Moishe* by the young people who served with him in the Israeli military.)

The first three days of my visit, as described in the trip report which follows, were devoted to visiting local factories that were again up and running, making wood tongue depressors. Ice cream sticks, and furniture and a textile plant that was planning to replace all of its water-damaged equipment with used equipment to be brought in from American firms who were upgrading their production lines.

When it was finally time for my talk on *Acciones Concretas y Avances en Comercio Electronico,* which was to close the conference, I was surrounded by eager young, potential industrial engineers, who seem to have absorbed every word like sponges. Best audience of my career. They all accepted the fact they were facing a new future of computers, electronic commerce and working as teams, using "appropriate" technology, rather than building fiefdoms, jealously guarding their results, when so much more could be achieved by working cooperatively. I left *Tegucigalpa* on a high, determined to do what I could to assist them rebuilding their ravaged housing and factories with the simplest, cheapest materials available.

Report on the Sights of Tegucigalpa, Honduras
by John Larry Baer, P.E.

"As an invitee to speak at UNITEC's Industrial Engineering conference, I thought I should take the opportunity to see how I might be able to help Honduran companies devastated by Hurricane Mitch. In addition I wanted to see if we could help Honduran industry to set up a facility to fabricate low cost building panels to quickly replace the homes of people left homeless in the extremely flood-prone Rio Grande riverbed.

It's not often that one receives a royal welcome, and thus the abrazo extended by my young hosts in the VIP lounge of this little airport, was not only unexpected, but a delightful surprise. After the initial cordialities, UNITEC's very young Dean of Industrial Engineering spotted the widow of the late and distinguished Mayor in the lounge, being interviewed. This charming lady, who has now stepped into her husband's shoes, became Jessica Calderon's immediate target to introduce this humble traveler and my planned mission to aid in the rebuilding of the Hurricane battered region. Madam Mayor, in typical, gracious Southern charm, invited us to a meeting in her office with her deputy responsible for the reconstruction.

(It should be noted that she has taken up the cudgel her late husband wielded until his untimely death in a helicopter crash while inspecting the hurricane damage-to get a law passed which would preclude

people from re-building in the extremely flood-prone Rio Grande riverbed. She is actively continuing his work of erecting modest shelters for the displaced, in a safe place, high on the hill overlooking their former homes.)

From such a moving beginning, I knew this was going to be a great trip, and, at the end of just one day, I felt that the people I met and the interest evoked in working together to rapidly build thousands of homes for hurricane displaced people, was well established. Antonio Fraile, the young Coordinador de Operaciones for Organizacion Internacional para las Migraciones, was very interested in the sandwich material with foam insulating core that I had brought along from World Building Systems to build some 10,000 6 x 4 meter homes.

The squatter shacks high up on the hill, well removed from the Rio Grande were inhabited by the displaced residents who had opted to return to the area. To see the location of these corrugated sheet and cardboard structures under the bridges, and in currently dry, open stretches, is to boggle the mind of anyone who can look just a few months ahead to even modest rains, that would likely, swiftly wash away their tarpaper hovels.

The provisioning of elemental housing for these hapless folk is the kind of manageable task that could provide quick results, unlike the Navy Seabees' longer term rebuilding of some of the 10 bridges ravaged and washed away by Hurricane Mitch, leaving intact only 2 bridges to connect Tegucigalpa with its neighboring Comayaguela. All these bits of wisdom were fed to me by Blanca Trinidad, a recent UNITEC graduate and charming chauffeur and raconteur, during an extraordinary day of touring and photographically recording the horrendous damage the storm wrought.

With the help of a young UNITEC I.E. graduate and the Director of ANDI (Asociacion Nacional de Industriales de Honduras), we visited the very small firm of Closet y Puertas where they fabricate modest metal cabinet doors and some wooden kitchen cabinets using salvaged and home-made sheet metal equipment. On a far more profound scale, we visited Fabrica de Cajas de Madera where men were trying to locate and then, using shovels, dig out totally mud immersed pieces of *equipment, ovens, and even file cabinets of ravaged, water and mud damaged paper files. In one corner, on top of the 3-meter layer of mud, the river had dumped a truck-through the roof of the factory–a sight to see and remember!*

We also visited Textiles Rio Lindo, where they are completely rebuilding their water treatment facility to be able to use the sewage-laden water from the Rio Grande. The region's major problem in my eyes is the restoration of bridges and roads. Traffic at the two remaining bridges and ALL roads leading to them, is a constant, nerve wracking hassle. Many of the roads leading to nearby factories are deeply rutted, with deep potholes and shoulder often totally washed out-making driving extremely hazardous.

I felt that all possible help needed to be extended by the American and European engineering world to help Honduran factories to rebuild and provide the good, used equipment they need to rebuild at the lowest cost and fastest pace. They have the will, but they need the means-and that includes information on recycling precious resources like their water, and how their workers could work more productively and safely.

Honduras' potential extends from the dedicated folks who own and run the factories to their most outstanding asset-their bright, intelligent, young people. The students and young graduates I met at the Universidad Tecnologica Centro-Americana (UNITEC) represent the greatest resource for this devastated country."

Congreso Sostenibles en la Industria de Alimentos, *San Jose, Costa Rica, Julio* 1999 Keynote

Five months after that momentous trip to Honduras, I was invited to give the keynote at the *1er Congreso Nacional de Ciencia y Tecnologia Sostenibles en la Industria de Alimentos*-The 1st National Congress of the Science & Sustainable Technology for the Food Industry.

The keynoter at the Plenary Session was supposed to be *Ing. José Maria Figueres,* the former *Presidente de la Republica de Costa Rica.*

As I sat on the dais with *Dottoresa Carmen Sanchéz, Directora Relaciones Institutionales de Ecoembalages España, Madrid, España,* and Dr. Robert Pagan, Director of "Clean Production" in the Food Industry, a UN program, from Brisbane, Australia, we chatted, awaiting *Ing. Figueres*. When word was received that he would be delayed, the conference organizer, *Ruth de la Asunción*, asked if I would be so good and give the opening talk. How could I refuse? Half way into my talk, the very young *ex-presidente*, strolled in, and, amicably suggested that I continue with my talk. Later, after he had given his speech, I was surprised to find that *Figueres* was quite fluent in virtually unaccented English–which he had learned at the university in the U.S. Other speakers from Brasil, Israel, Mexico, Spain, Costa Rica and the U.S., as well as the Secretary General of *Federación Española de Alimentación y Bebidas,* followed.

But they weren't through with me yet. At the closing ceremony, *Clausura y entrega de premio ASCOTA,* Ruth de la Asunción and the Dean. *Gisela Kopper,* came over and knelt by the seats where Shirley and I were sitting and asked if I would be so kind and to provide some closing remarks and wrap up the conference. Well, while the entertainment was going on in the auditorium, I did some fast and furious scribbling, writing down some key points for a closer. Happily, it all went well, and after the evening's splendid supper and dance, we prepared for a day's outing, seeing the rain forest from an aerial tram on an overhead gondola, watching a sloth slothfully crossing the road on an overhead power cable, and trying to catch a glimpse of the frequently shrouded *Poas* volcano.

1er Congreso Nacional de Ciencia y Tecnología de Alimentos, San Jose, Costa Rica, 19 julio de 1999

1999 was to be a wonderful and busy year. Early in the year I had received an invitation from the conference *Coordinadora*, Ruth De La Asuncíon to be a Plenary speaker on *"Acciones concretas para le producción de alimentos."* (concrete actions to improve the production of food products.)

I was actually to be preceded by Ing. José Maria Figueres, the former president of the Republic of Costa Rica, who was to provide the keynote. As luck would have it, he was late, and so yours-truly was asked to give the key-note. How could I refuse? As it happened, he showed up shortly after I had begun my talk, but graciously waited for me to finish, before he gave his talk.

Later we had the opportunity to chat amiably, as he was totally fluent in English. My opening was as follows:

Señoras y Señores
It is a great pleasure for me to be here to speak to you today.
And more than that, it is a very great honor to speak after your very distinguished president, Engineer Figuera.
Es un gran placer/ para mí/ estar aquí/ para hablarle hoy. Y más de eso,/ es un muy gran honor/ para hablar/ después de su presidente/ muy distinguido,/ Ingeniero Figuera.
Please forgive me if I speak to you in English, but my Spanish is poor and I have much new information to bring to you.
Por favor/ perdóneme/ si yo le hablo/ en inglés,/ pero mi español/ es pobre// y yo tengo/ la muy nueva información/ para traer a usted.
Since I have much more information than I could possible deliver in twenty minutes, I will make myself available to you during the conference in case I can be of further assistance to you.
Desde/ que yo tengo mucho más información/ que que yo pude posible/ entregue en veinte minutos,// yo me haré disponible/ a usted/ durante la conferencia// en caso de que/ yo puedo/ ser de ayuda extensa/ a usted//.

(I left my / and // spacers in this copy, because I found it to be an excellent pacing tool–especially when speaking in a foreign language.)

A GRATEFUL REFUGEE KID'S RECOLLECTIONS

Thoughts on Costa Rica

Oh tico tico tic, oh tico tico tac, oh tico tico tico all the day. That was Carmen Miranda's theme song in her day (and ours), when she sang with Xavier Cougat's orchestra. And while it may not be the theme song of today's Costa Rica-they still operate on "tico time." Not always right on the minute-but overall, both the *Congreso Tecnologias Sostenibles en la Industria de Alimentos,* and our tours were remarkably "on time." And the people-every one, charming and helpful; and we met some really interesting and wonderful folks along the way.

We arrived in San Jose, after almost an hour's sitting on the Dulles runway waiting to take off, and a very brief stopover in Mexico City, well after midnight. But there was our tired, but as ever charming hostess, Ruth de la Asuncion Romero, waiting at the gate, to whisk us through passport and customs checks and off to our sumptuous *Herradura* Hotel, just 10 minutes away. When you are the deputy director for a large conference, you'll know what this kind of tired is.

Sunday, after a leisurely "breakfast at Tiffany's" (that's the clever name of their dining room) I made it my first business to load my Powerpoint presentation from the slides in my pocket to their computer. And this was the first of many exposures to the wonderful helpfulness of the University of Costa Rica staff. Between English, French and Spanish we managed to communicate and load the precious data, including the six slides added the day I left, with the very latest news from this week's Business Week, Information Week and Integrated Solutions.

Duty done we took the hotel's free shuttle bus into San Jose, some 20 minutes away. And spotted a huge Menorah in front of their Arts Museum! We learned later that Costa Rica's 1st president, the current lady-VP and many senior government and other influential people, were Jewish. San Jose appears to be a "modest" town, akin to some of our smaller Mid-West cities. It has many, many parks, coffee plantations wherever you look and a population of just 3.5 million, with 1 million, now legal, Nicaraguans, who toil in the fields.

Monday, the Congreso got off to a fine start, with a moving address by Dr. Owen Fennema, the head of the Industry of Food Technology. After a short coffee break, it was time for the 1st plenary; but our keynoter, *Inginiero* Jose Maria Figueres Olsen, was late-sooo, Papa Baer into the breach. Despite having 64 slides, I got through most of them when my cake timer went off after 19 minutes, and I skipped just a few of the late additions, to finish in 21 minutes. Since I had brought along 70 copies of my slides, which went like hotcakes, I didn't feel too badly.

Then the fun began. Dra. Carmen Sanches from Spain, talked about Eco-packaging for FORTY minutes, despite the repeated, pleas of our very frustrated chairlady to get her to finish her meandering diatribe. There is always one like this in every crowd. Dr. Robert Pagan, UNEP Australia, gave a fine talk about ecological measures and the need for cleaner production in food processing, which is traditionally one of the largest industrial sectors in any country. Finally, the ex-*presidente*, gave a fabulous, spell-binding talk about Costa Rica's achievements and future plans on sustainable food production. BTW, when you see long names, the next to last is the father's name and the last name is the mother's maiden name–as in Figueres Olsen.

In the afternoons we goofed off, lazed by the terraced pools and readied ourselves for the Welcoming Poolside Cocktail Party Monday, the Concierto by *Quatro Trombones* Costa Rica at their grand, old concert hall on Tuesday, and the farewell Party with noisy, lively dance music and *tipico* BBQ at the *Campestre Espanol* out in the country, on Wednesday night.

The other four plenary sessions provided excellent presentations on Food Safety & Quality, better Product Standards, Innovations and Product Trends, and Productivity & Competitiveness in Small & Medium Industries–critical for countries like Costa Rica. Among the speakers was Dr. Al Clausi, the distinguished retired VP of General Foods, who provided a great background on New Food Development–ideas that succeeded, like Tang & Dog Burgers–and some that failed. Dr. Jorge Jordana, President of FIAB in Spain, talked about "functional food" and the fact that "what you eat determines your state of health."

Dr. Franco Lajolo from Sao Paulo, Brazil, reminded us that 'people want to know what they should accept in genetically modified food.' He also noted that there are new foods, which are easier for diabetics to digest & reduce blood sugar. Oscar Harassic from OAS, advised PYMEs that they need to team up for strength. *(Pequeno Y Medio Empresas)* Dr. Yoram Malevski, also from OAS, spoke about "the need for discipline and cleanliness in food processing–as the Germans learned from their mothers." Daniel Vilarino, OAS Environment Specialist noted the value of recycled material–Reuse/Reduce and Recycle–an underlying theme throughout the Congreso.

Dr. Arnold Ventura, President of CARCYT, Jamaica observed that "big US farm distributors say "We deal in bulk–not in discretion" and that the US Dept. of Agriculture refuses to segregate genetically modified (GM) from non-GM foods–a hint that Agribusiness is driven by greed, that they waste water, and deny customers the knowledge they need to choose healthy foods wisely. In this regard one of the reasons for the high cost of foods in Japan, is that they insist on segregating organically grown food, which has raised some food prices three-fold. It was also noted that Germans have the highest consumption of ecological foods & that children with high consumption of beans are likely future candidates for diabetes.

Tuesday night's wonderful, fun, concert was followed by a grand VIP dinner at the Club. It was only when we got back to our room, that I noticed that our invitation was addressed to: John Baer, Guest of Honor. I was touched–as I was also, when Gisela Kopper, the director, and my friend, Ruth, who is her deputy, stopped by our seats at the Concierto, & asked if I would please lead the closing ceremony, as they considered me "part of the family." Who could refuse such a warm invitation? We cancelled our tour of the Butterfly Farm; I sat down to compose some warm, heartfelt, remarks. Fed them into the PC, my good sidekick edited them, and we ended the Congreso on a happy note.

Thursday we had a 5:30 wake-up call for our 6:30 bus to the *Poas* Volcano and Cloud Forest (being *Inverno* or winter in Costa Rica, it was shrouded in fog, so we had to settle for the picture of the lake.) On the way to a *tipico* country breakfast we watched a sloth, moving slothfully, upside down on the power cable along the road, gingerly moving one paw at a time, till he was comfortable & safely on an adjacent tree branch, and disappeared into the "living fence" protecting the adjacent coffee plantation.

From there we just managed to cross a rickety, old wooden bridge at the imposing La Paz Waterfall. After our bus crossed, they closed the bridge for four hours for repairs! Whew. On to lunch and then a very wonderful highlight of our trip-a quiet ride of the *Sarapiqui* River. The kids would have loved it. We saw white-faced monkeys, great herons, calmly spreading their wings to dry on riverside logs, Crocodiles and Iguanas sunning themselves on the water's edge (didn't get too close to the Croc, because one swipe of his tail could have pierced the side of our boat.)

Didn't see any *Oropendulas*-maybe next trip. And then a drive on a fine highway through the huge *Braulio Carrilio* National Park back to the *Herradura* Hotel. The road took us through a tunnel cut through the mountain range that runs down the length of Costa Rica, built to expedite the flow of trucks which carry the coffee, pineapples and bananas for export to Caribbean and Pacific ports.

After the 11 hour trip Thursday, we had one more thrill in store-getting up at 5 for a 5:30 bus to the 1000 acre Rain Forest & a tour on their Aerial Tram. Whew-talk about puckering. After another country breakfast at a lovely Swiss-style chalet, we boarded a wire-caged tram, that took us on a 90 minute trip to 20+ meters through the middle of the forest, allowing a very close look at the foliage, with a trained guide providing detailed explanations. Then on the return leg the tram rose to some 50 meters! to see the top of the canopy. A real thrill & delight.

We ended the day with a scrumptious dinner on the balcony at the nearby grand Marriott Hotel, with our charming hostess, Ruth. Then off to bed for a 4:00 AM wake-up to catch our 6 AM, 8-hour flight home. I must say, to UAL's credit-we breezed through the check-in, delightful crew service, and a relaxing-albeit rather long and uneventful flight home. It was a great learning experience in a new field. Food = Health! *Alimentos = Salud y dineros!*

Helping Siemens transition into Y2K–July 1999

Well, that was almost my last great adventure as a consulting engineer, but there was one more, grand finale. Another one of my industrial engineer buddies called me and asked if I could come to Germany for a few weeks to help Halifax Corp. assist *Siemens* assure a safe transition into the next century. They needed a "contingency" or "emergency" plan for 1.1.2000. On very short notice I flew to Dresden, which was formerly part of East Germany, and had been bombed heavily by our American planes.

Once I got there and was introduced to the working staff I realized that, in typical German fashion, they had everything under control and didn't need help from these outsiders. Well, it happened that Siemens had some $ 7 million worth of computer chip inventory in process on 20,000 pieces of equipment, on any given day and a daily production of some $ 3 million. And there was a communication problem between the German staff and the Halifax team. So, the first thing I did was to meet with each of the managers–in production, design, financial, etc., and explained to them *auf Deutsch* (in German) that if they didn't cooperate and work with us, they might be out of work come Y2K! It worked!

In short order the team contacted al the suppliers and the manufacturer of each piece of equipment to get their written verification that their equipment was Y2K capable. There were just 14 vendors in both facilities and production for whom we had a problem getting certification. One had the gall to say, not to worry, that they would get their equipment Y2K capable in time. We wrote back to them, quoting the head of General Electric: *"comply, or die."* They quickly came around. We completed an initial audit in weeks 36 to 39 in September, and proposed a "final audit" for weeks 48 to 51.

That year, New Year's Eve 1999, was celebrated at Siemen's conference center, with all top management present, and not even a hint of wine or liquor. It was not until Midnight chimed, and there was no break in production, that they broke out the champagne.

I'd like to add, in passing, that while touring the famous Dresden cathedral, our tour guide chastised us, saying that all this destruction "vas bekauz of vot you Amerikans did." To which I answered her in my best German "Lady, we didn't start the war–you did." It was a fitting climax to my career.

Part 8
Conclusions and "Lessons Learned"

It's been a good, rich and rewarding life in my adopted country, which gave me life and a career. I've learned to always hire people smarter than me so that they can help me with the challenges assigned to me. I've learned that retirement presents a wonderful opportunity to volunteer, to help others who were not as fortunate as I was. This 3rd career has provided me some of the best hours of my life, helping children to learn to read, immigrants to learn English, or pushing wheelchairs. I don't ever want to just sit back and loaf–that's not me. I want to spend the rest of my life trying to be useful every hour of the day. It's a blessing to be able to do that–and to enjoy our children and grandchildren, and, above all, to enjoy every hour of the day with my wonderful wife.

PUBLICATIONS:

Editorial Review Board & Contributing author on Manufacturing Technology Handbook (1980), Flexible Machining Systems Manual (1981), Military Handbook No. 727-Design Guidance for Producibility (1984), Critical Technology Classification Guide (1981 & 82), "Foreign Dependency in Military Purchasing" (1986), "Improving Productivity Through Manufacturing Technology (1980), "The Other Side of Outside Sourcing" (1985), "Foundry 2000, a Technological Forecast" (1984 contributing author), "Planning & Budgetary Control Applied to R&D Projects"(1955), "The Navy's Anti-Tank System" (1989) and author of over 50 other reports, papers & journal articles.

TECHNICAL PRESENTATIONS:

Lecturer to the CIA, Defense Systems Management College, Army Logistics Management Center, Project Managers Orientation Course, Foreign Science & Technology Center, ADPA Conferences, Industrial Preparedness Conferences & Manufacturing Technology Advisory Group first Foreign Technology Seminar initiator & chairman, Westchester Community College & numerous clubs & engineering organizations (1960 to present).

Worked on CALS Conference 1990 to 1997, incl. as Plenary Chair & International Chair

Managed & arranged for speakers to ilce '93 & '94, Montpellier & ilce '95 & '96 in Paris, France

Tutorial & Seminar speaker on Computer Aided Acquisition Logistics Support & Electronic Commerce: Pacific CALS Conference, Seoul, Korea, Sept 1996; 12th International Logistics Conference, Glyfada, Greece, Sept 1996

Electronic Commerce Applications for Swiss Telecom, Berne, Switzerland, Oct. 1996

Japanese Delegation to Intl CALS Conference, LongBeach, CA, Oct 1996

Electronic Commerce Implementation for Japan Institute of Office Automation, Mar. 1997

Korea Institute of Construction Technology in Washington, DC, Mar. 1997

Chair & Keynote Speaker of 1st Mediterranean CALS Conference, Istanbul, Turkey, June 1997

CALS for the Verbindungstelle des Deutschen Ruestungsbereiches, Reston, Va Aug. 1997

Electronic Commerce World Conference, Philadelphia, PA, Sept. 1997

CALS Europe Conference & Expo, in Frankfurt, Germany, Oct. 1997

Library Information Technology Conference, Rio de Janeiro, Brazil Nov 1997

Society of Logistics Engineers Seminar, Athens, Greece, June 1998

Hellenic Federation of Information Enterprises' "Money Show," Athens, Greece, June 1998

Keynote Conferencista Simposium de Ingenieria Industrial, Tegucigalpa, Honduras, Febrero 1999

Keynote 1er Congreso Nacional de Ciencia y Tecnologia de Alimentos, Costa Rica, julio 1999

An Abbreviated Timeline for my generation

1933-Jan 30-Adolf Hitler becomes Chancellor of Germany.

March 12-First concentration camp opened at Oranienburg outside Berlin.

April 1-Nazi boycott of Jewish owned shops. **May 10**-Nazis burn books in Germany.

In June-Nazis open Dachau concentration camp.

1934-June 30-The "Night of the Long Knives."

Aug 19-Adolf Hitler becomes Führer of Germany.

1935-Sept 15-German Jews stripped of rights by Nuremberg Race Laws.

1936-Feb 10-The German Gestapo is placed above the law.

1938-March 12/13-Germany announces 'Anschluss' (union) with Austria.

Nov 9/10-Kristallnacht-The Night of Broken Glass.

1939-Jan 30, 1939-Hitler threatens Jews during Reichstag speech.

In Oct-Nazis begin euthanasia on sick and disabled in Germany.

1941-In June-Nazi SS Einsatzgruppen begin mass murder.

Sept 1, 1941-Nazis order Jews to wear yellow stars.

Sept 3, 1941-First experimental use of gas chambers at Auschwitz.

Sept 29, 1941-Nazis murder 33,771 Jews at Kiev.

1942-Jan 20, 1942-SS Leader Heydrich Conference on "Final Solution of the Jewish Question." **June**-Mass murder of Jews by gassing begins at Auschwitz.

July 22, 1942-First deportations from the Warsaw Ghetto to concentration camps; Treblinka extermination camp opened.

Oct 5, 1942-A German eyewitness observes SS mass murder.

Dec 17, 1942-British Foreign Secretary Eden tells the British House of Commons of mass executions of Jews by Nazis; U.S. declares those crimes will be avenged.

1943 **April 19, 1943**-Waffen SS attacks Jewish resistance in the Warsaw ghetto.

May 16, 1943-Jewish resistance in the Warsaw ghetto ends.

June 11, 1943-Himmler orders the liquidation of all Jewish ghettos in Poland.

1944-**Aug 4, 1944**-Anne Frank and family arrested by the Gestapo in Amsterdam, Holland.

Oct 30, 1944-Last use of gas chambers at Auschwitz.

1945-**April 29, 1945**-U.S. 7th Army liberates Dachau.

REFERENCE MATERIAL & ACRONYMS:

CD ROM = Compact Disc-Read Only Memory
CIO = Chief Information Officer of a company
EC = Electronic Commerce
EDI = Electronic Data Interchange
EDIFACT = Electronic Data Interchange f/Administration, Commerce & Transport
FTP = File Transfer Protocol
http = hypertext transfer protocol
ISDN = Integrated Services Digital Network
CKO = Chief Knowledge Officer of a corporation
NC = Network Computer (stripped down, basic Personal Computer or PC)
OCLC = Online Computer Library Catalog
PoP = Points of Presence (commercial market depth)
SGML = Standardized Graphics Mark-up Language

I was going to include a copy of the last presentation I made, on:
"Electronic Commerce, Application Standards & Internet-Key Elements of the Information Society"
by *John Larry Baer, P.E., President, International Management & Engineering Consultants*
  An Introduction to an Engineer's View of the Information Society and its Relevance to the Library System
which started as follows:

Senhoras e senhores:

Muito prazer em conoce-lo. I feel a bit like Dom Pedro Alvares Cabral, the Portuguese explorer, who discovered this great country in 1500. Of course, the landing approach is a little different today, nearly 500 years later. I am an engineer, not an expert in library science like you. But I have been a lifelong USER of libraries and an avid reader all my life. I learned the Dewey decimal system for cataloguing books at about the same time that I learned English back in the early 1940s. You see, my first job in the United States was working in the stacks at the Brooklyn Grand Army Plaza Library. And my local school librarian tells me that they still prefer the Dewey Decimal system as being much easier to administer and for their students to find things.

Many things have changed since then in terms of the accessibility of INFORMATION. Granted, we are striving to become a paperless society to improve the way we do business and, not least, to save millions of trees in the process. But it's difficult to work on a laptop computer while riding on the Metro, or sitting in the bathtub or especially to catch up on your bedtime reading. So-take heart, we are not likely ever to completely do away with paper. A love note left on the computer for your wife to read is just not the same thing as a personal note left on her pillow.

But why am I, a simple engineer speaking to all of you enlightened library scientists? Because you need to know what I need from you. I am your customer! and these days the company which fails to go out of its way to delight its customers with swift, attentive service, quickly goes out of business.

But decided against it. However, if you're interested in this topic, just send me an e-mail to jsbaer@verizon.net, and I'll send you a copy.